D1345856

THE TEACH YOURSELF BOOKS
EDITED BY LEONARD CUTTS

ASTRONOMY

TEACH YOURSELF

ASTRONOMY

By

DAVID S. EVANS

Chief Assistant, Royal Observatory
Cape of Good Hope

Illustrated by the
author

THE ENGLISH UNIVERSITIES PRESS LTD
102 NEWGATE STREET
LONDON, E.C.1

First printed 1952
This impression 1963

Made and Printed in Great Britain for the English Universities Press, Ltd., London by C. Tinling & Co., Ltd., Liverpool, London and Prescot.

CONTENTS

For Katie, Cyril and Betty
in gratitude.

LIST OF PLATES

Inserted between pages 86 *and* 87.

INTRODUCTION

Essential Preliminary

I myself rarely read introductions: this one is addressed to the reader, with an aside to the reviewer, and he is asked to read it to avoid misunderstanding. The title of this book is *Teach Yourself Astronomy* and, as far as possible, I have stuck to my brief. Although it is one of the essential principles of science that all scientific knowledge can be recaptured by anyone, given sufficient time, intelligence and patience, life is too short to make this a practicable course. I have therefore often been compelled to set out the details of book knowledge of astronomy. Apart from this, my principal problem has been to try to make this book a means of elementary self-instruction and yet, at the same time, to make it up-to-date. The reader must judge for himself how far this object has been achieved.

I have assumed that the readers of this book will already have decided for themselves that a knowledge of astronomy is a desirable thing, and worth the exertion of some mental effort to acquire. I have not therefore been at pains to let the reader down particularly lightly. The best way of reading the book is probably to run through it quickly at first to see what it contains, and then to study it in more detail. I have not had much recourse to mathematics, at least formally, but at various points there will be found trains of reasoning which are mathematical in nature. This choice is deliberate, because many people are, as it were, slightly allergic to mathematics, but can digest the equivalent verbal reasoning without discomfort.

It is impossible to include all astronomy in the compass of such a book and, for example, there are only slight oblique references to astrophysics, and, deliberately, nothing at all about precession, nutation or aberration. My professional colleagues may hold up their hands in horror at the omissions, and at the superficial treatment of many of the topics discussed. But the book is not written for them; it is written for the interested man in the street, in the hope that it will give him pleasure, and that a few of his number will be sufficiently interested to go further, and eventually to join the ranks of one of the various astronomical associations as serious amateur observers.

CHAPTER I

FUNDAMENTALS

This book is written on the assumption that you know no astronomy, and you may believe that this is strictly true. In fact, however, almost everybody does know some astronomy and we can best begin by putting down what everybody does know and make this our starting point. Everybody knows that the earth turns on its axis from west to east, so that the sun, moon and stars go past us in the reverse direction from east to west. Everybody knows at least one constellation: in the northern sky the Plough, in the southern, the Southern Cross; and everybody knows, but may not realise the fact until it is pointed out, that these constellations, and all others, retain the same shape so far as observation with the naked eye can show.

Now go out of doors on a clear dark night and take note of what you see. About half an hour will be sufficient to show a number of important facts. If you watch carefully you will see that the stars, like the sun and the moon, rise in the east and set in the west and that all the stars and constellations keep station relative to one another. The sky looks to the observer like a huge hemisphere with the stars marking particular positions on the inside of it. The horizon forms the lower rim of the hemisphere, and the point directly over the observer's head is known as the *zenith*. We can, more conveniently, think of the whole sky as forming a complete sphere, of which the lower half is cut off by the horizon.

The Celestial Sphere

This imaginary sphere, with the observer at its centre, is called the *celestial sphere*. The sky is not, of course, a real sphere, but the idea of a celestial sphere, that is a large sphere with the observer at the centre, and the positions of the various stars marked on its inner surface, is an extremely useful one, and will be used throughout the sequel. A physical realisation of the celestial sphere is afforded by the planetarium, a darkened building with a white domed ceiling, on to which light-spots in positions corresponding to stars and planets can be projected by an optical device. These spots can be moved so as to reproduce exactly the motions of the sun, moon, planets and stars, and their rate of movement can be increased so as to demonstrate phenomena occupying very long periods in nature. The illusion of being out of doors on a clear starry night is complete. Although the planetarium was first devised in Germany, almost all planetaria open for public exhibition are located in the U.S.A.

The celestial sphere appears to be turning from east to west in consequence of the rotation of the earth from west to east. Now, when a globe turns on its axis there are two points—the points where the axis goes through the surface of the globe—which do not move at all. In the case of the celestial sphere these two points are called the north and south *celestial poles*. The Pole Star, which is located by the two

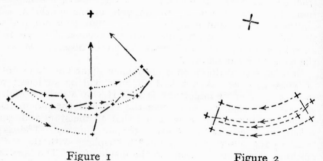

| Figure 1 | Figure 2 |

pointers of the Plough is very close to the true position of the north celestial pole. If you live in the northern hemisphere you can easily see, by drawing a little sketch every hour for several successive hours on one night, that although the Plough appears to turn round the north pole of the heavens, the pole star remains almost exactly fixed in position. (Figure 1.)

It is a mere coincidence that there happens to be a bright star so close to the north celestial pole. In the southern sky there is no bright star close to the south celestial pole, but, by observing the movement of the Southern Cross for several hours it is possible to verify that there is also a fixed point, the south celestial pole, in this half of the sky. (Figure 2.)

Just as the earth has a north and a south pole with an imaginary line, the equator, half-way between them, so the sky has a *celestial equator* which is half-way between the north and south celestial poles.

Now, let us look at matters from a slightly different standpoint. In Figure 3 let the circle represent a section through the centre of the earth with an observer standing, looking at the sky, from some place in a moderate northerly latitude. We denote the latitude of a place on the earth by the Greek

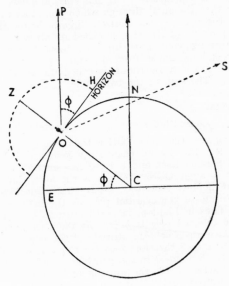

Figure 3

letter φ (phi). We therefore mark the position of the observer on the surface of the earth in such a way that the angle OCE is equal to φ. The height of the observer compared with the radius of the earth is supposed to be negligible. If we draw a line from his eye to the pole star, or more precisely to the true position of the celestial pole, this line will be parallel to the axis of the earth. If we try to imagine the sky as the observer sees it, we shall first have to mark a line touching the surface of the earth, which represents his horizon.

We can easily see, by drawing a dotted line to any star, S, which, in the figure lies below this horizon, that such a star will be hidden by the bulge of the earth. Sketching in the celestial sphere for the observer at the given position we mark his zenith at the point Z, and the visible portion of his celestial sphere as the semicircle above the line representing his horizon. The angle between the direction to a star and the horizontal is called the *altitude* of a star. If, on looking at the sky you choose any particular star, and face directly towards it, then the angle through which you must turn your gaze from the point of the horizon directly below the star, up to the position of the star, is called its altitude. If you continue turning your gaze upwards towards the zenith, then the remaining angle, from the star, up to the position of the zenith, exactly overhead, is called the *zenith distance* of the star. The sum of the two angles is of course ninety degrees or,

For any star

altitude + zenith distance = 90° (one right angle).

In the diagram, the angle POH is the altitude of the north celestial pole. But by an elementary geometrical theorem about two parallel lines and a third line cutting them, the angle ZOP = angle OCN, where N is the north pole of the earth. But angle OCN = 90° − φ so that angle POH = φ, i.e. the altitude of the celestial pole = the latitude of the observer. Thus in London, in north latitude 51½°, the altitude of the celestial pole is 51½°. At the north pole itself, where the latitude is 90°, the altitude of the pole is 90°, and its zenith distance is therefore zero; or to put it another way, the pole of the heavens is in the zenith as seen from the north pole of the earth. From the equator, at latitude 0° the altitude of the pole is zero. That is, if you are on a ship, on the day you cross the line, the pole will just lie on the horizon.

If you follow through the foregoing arguments it will easily be seen that precisely the same arguments could have been used in the case of the south pole of the heavens. In south latitudes φ degrees the south pole will be φ degrees above the horizon; at the south pole of the earth, the south celestial pole will be directly overhead.

Navigation by the Stars: The Sextant

This simple rule enables us almost at once to make a practical application. Clearly the navigator of a ship will find it extremely easy to determine the latitude of his ship on any night when he can see the pole star. Unfortunately there is no star exactly at the pole, but with a quite small error, not small enough unfortunately to be negligible in practical navigation, he can, by using a sextant to measure the altitude of the pole star above the horizon, determine at once what the latitude of his ship is.

The principle of the sextant is illustrated in Figure 4. M and M′ are two plane mirrors, the latter being fixed to the frame of the instrument while the former is mounted on a rotating arm which carries the pointer P moving over the graduated arc A. T is a telescope through which the operator

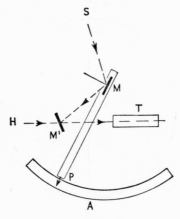

Figure 4

looks. When he does so he sees two images. One is an image of the horizon brought to him along the horizontal light ray H which passes through the mirror M′, this being only lightly or half-silvered so as to transmit light as well as reflect it. In the second place he sees the image of the sun, moon, planet or star by means of the light ray from S which is reflected at M, then at M′, and then into the telescope T. This will only occur if M is at the correct angle. As P is moved over the arc A at one moment the operator will see the sun in his field of view exactly in line with the horizon. He then reads the angle on the arc A. In fact if the sun were to move up through a certain angle the mirror M would have to be turned through only half this angle because of the well-known property that turning a mirror through a certain angle turns the reflected ray through twice that angle. However, to save trouble, the arc A is graduated in half degrees so that the observer can read off the altitude directly. When M and M' are parallel the operator sees the horizon both directly through M' and twice reflected in M and M'. The zero point of the graduation of A thus corresponds to a point on the right of A such that PM is then parallel to the face of M'. The graduation in half degrees starts from this point.

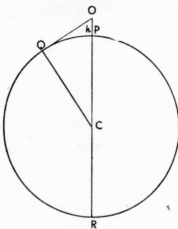

Figure 5

Sextants are usually beautifully made and include all sorts of refinements such as verniers for accurate reading and dark shades to protect the observer's eye when " shooting the sun ". Sextants for use in aircraft, when the horizon is invisible, also contain a bubble arrangement which shows when the sextant is being held level. This does duty instead of the observed horizontal HT but is less accurate, so bubble sextants often contain devices whereby many altitude readings of the same star—up to 100—may be taken very rapidly and the average value shown automatically. This cuts down the greater errors inevitable in the process of attempting to make accurate measures from a rapidly moving and unsteady aircraft. The accuracy obtainable with a first-class marine sextant is about $\frac{1}{360}$ of a degree and with a bubble sextant about $\frac{1}{60}$ of a degree.

The Distance of the Horizon

In discussing the rule that the altitude of the pole was equal to the latitude we made the proviso that the height of the observer was negligible. When this is not the case, as happens in observations from a high deck, a mast or an aircraft, a correction must be applied. In Figure 5 let O be an observer at a height h above the surface of the earth. The line OQ just touches the earth and is the direction to the horizon as seen by O,

Now by Pythagoras' theorem

$$OQ^2 = OC^2 - QC^2 = OC^2 - PC^2$$

for QC = PC = radius of the earth.

But $OC^2 - PC^2 = (OC - PC) \times (OC + PC)$
$$= OP \times OR$$

The distance OR = diameter of the earth (neglecting the small extra height OP).

Hence $OQ^2 =$ (distance to horizon)^2 = height of observer × diameter of earth.

Taking the earth's diameter in round figures as 8000 miles then
distance to horizon in miles = $\sqrt{\text{height of observer in miles} \times 8000}$

$$= \sqrt{\tfrac{8}{5} \times \text{height of observer in feet}}$$

approximately.

Also, the true horizontal through O is perpendicular to OC. Hence Q is too low, i.e. the horizon is depressed, as the result of the elevation of O above the surface. This angle of depression is equal to the angle QCO (note that this is zero when OP is zero, as it should be). It can easily be shown that roughly angle QCO = $\sqrt{h}$ minutes of arc where h is in feet. One minute of arc (1') is one sixtieth of a degree. This correction must be subtracted from the observed altitude of any body as measured with a sextant to obtain the true altitude. This corrected altitude is the value to be subtracted from 90° to obtain the true zenith distance. (See page 4.)

The rule shows that a six-foot man with his eye 5 ft. 6 inches above the surface can see 3 miles. From a ship's mast 100 feet high the visibility is about 13 miles. From an aircraft at 10,000 feet one can see about 130 miles. The depression of the horizon in these cases is respectively about $2\frac{1}{2}'$, about 10', and about $1\frac{1}{2}°$.

These distances must all be somewhat increased because of the bending of light rays by the atmosphere of the earth, an effect which is discussed in Chapter V.

Substellar Points and Position Circles

In discussing the rule that the altitude of the pole was equal to the latitude we remarked that at the north pole itself the pole of the heavens was directly overhead. Now, for each star there is some position on the surface of the globe at each moment, such that the star is directly overhead. This point we call the *substellar point* of that star. For the north pole of the heavens, imagining for the moment that we mark it by a star, the substellar point is the north pole of the earth, and is fixed on the earth. All the other substellar points are continually moving over the earth's surface as the earth turns on its axis. Now, our rule about the north pole can be put in a slightly different way. The rule that altitude of the pole is equal to the latitude of the place of observation can be put alternatively as the rule that the zenith distance of the pole of the heavens is equal to (90° — latitude) (We have merely subtracted both the altitude and the latitude from 90°). But (90° — latitude) is the angle from the pole to the position of the observer, or, what is the same thing, the angle measured at the centre of the earth from the substellar point of the pole to the position of the observer.

We need only redraw our diagram (Figure 3) taking any star whatever instead of the pole, P, to see that the angle measured at the centre of the earth from the substellar point of a star to the position of the observer, is equal to the zenith distance of that star as measured by the observer. This is the fundamental rule of all processes of star navigation. Now, to what distance on the surface of the earth does an angle of, say, one degree at the centre of the earth, correspond? The radius of the earth is about 3960 miles, so that the whole circumference of the earth is 2π times this, or 24,900 miles. This corresponds to an angle of 360 degrees at the centre of the earth, so that one degree corresponds to about 66 miles. In all science the degree is divided into 60 parts called minutes of arc, one of which is written as 1′. Each minute of arc is divided into 60 parts called seconds of arc, written as 1″. An angle of one minute of arc corresponds to a distance of 6080 feet at the surface of the earth, a length which is called a nautical mile. A speed of one nautical mile per hour is called a knot. Just to fix the ideas, a fairly fast ocean-going liner may do about 25 knots, corresponding to 600 nautical miles in one day, or, if the ship is travelling due north or south, to a change in latitude of 10 degrees.

The use of this sytem of nautical miles simplifies matters a great deal. If a star is observed to have an altitude of 80° then its zenith distance is 10°, corresponding to 600 minutes of arc. Our rule then shows us that the ship is 600 nautical miles away from the substellar point of the star observed. All one has to do is to convert the angle measured in degrees into minutes by multiplying by 60 and the result is the dis-

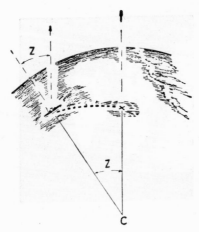

Figure 6

tance in nautical miles from the substellar point of the star observed. (Figure 6.) So if we know that we are, say 600 nautical miles from the substellar point of one star, and 1200 from the substellar point of another, all we should have to do to fix the position of our ship is to draw two circles on the globe with the correct centres and radii and then we know that we must be at one or other of the intersections. We thus need two separate observations of stars to fix the positions of our ship. Even then there is of course an ambiguity since the two circles—known as *position circles*—intersect in two points, but in practice the two intersections will be far apart—one may be on land or in a different ocean from the first—and the question of any real uncertainty does not arise.   (Figure 7.)

If we know the location at the moment of observation of the substellar point of the star observed, then this single observation tells us that the ship lies on a circle centred at the substellar point, and having a radius corresponding to the observed zenith distance of the star.   The case of the pole star, or the true pole, with which we began, is just an example of this rule.   The altitude of the pole is equal to the latitude, that is, the zenith distance is $(90° - \text{latitude})$.

Figure 7

The substellar point of the north celestial pole is the north pole of the earth.   Hence position circles for the pole star are approximately parallels of latitude on the earth, since these are lines on which one is at a certain fixed distance from the north pole of the earth.   The particular parallel is the one which gives the correct value of the zenith distance.

All this seems very simple.   All one has to have for navigation is a sextant to enable one to measure the zenith distance of any star one chooses, and then one has to have some way of knowing where the substellar point of this star is at the moment of observation.   Naturally, but unfortunately, these substellar points move as the earth rotates.   Think of a star such as for example the middle star of the handle of the

Plough, Mizar, which in Britain is almost dead overhead at midnight on a spring night. This means that at this moment the substellar point of Mizar lies somewhere in Britain. But at a later or earlier hour or in the middle of an autumn night, Mizar may be down near the horizon so that obviously then the substellar point of Mizar is somewhere completely different. Clearly the matter is somewhat more complicated than we thought, and we cannot, as we might if the earth were not rotating, simply arm ourselves with a book giving a table of the latitude and longitude of the substellar points of all stars.

Locating the Stars

In order to pursue this subject we must try to introduce some order into our method of defining star positions. We do this by means of numbers, or co-ordinates, or map references,—call them what you will—which resemble closely the similar numbers, latitude and longitude, which define the location of a place on the surface of the earth. The topography of the sky is as complicated as the topography of the surface of the earth. On the surface of the globe we need something more general and precise than this sort of recipe " Go 500 miles south from the Azores, turn left . . ." and so on. This sounds ridiculous, but in fact, in the sky, amateurs are liable to speak in just this kind of way ; " You see that bright blue star up there: yes, where I'm pointing. Now come down a yard to your right and you'll see a tiny little star . . ." and so on. Obviously we must have a bit more system than this.

I mentioned the phrase " a yard to your right " because that is the kind of thing people say. It is in fact a meaningless phrase, because when they imagine a yardstick being held up to the sky to serve as a measure of distance between stars, people do not say whether the yardstick is being held up at arm's length or at a distance of a hundred miles from the eye. A yard at these two distances will present an entirely different appearance. In defining positions of stars we do not speak of the real distances between them: all the stars are indefinitely remote and the real distances between them tremendous. What we have to do is to revert to our picture of a celestial sphere and to specify in which directions the stars appear. The celestial sphere, which represents the appearance of the sky with the eye of the observer at the centre, may be thought of as being of any size we like. But imagine such a sphere drawn, and imagine lines from your eye at the centre of the sphere drawn to any two stars. Then we can most conveniently say how far apart the stars seem to be by giving the angle measured at the eye between the directions to the given pair of stars. It will help if we give a few sample angles to fix the ideas. From the zenith down to the horizon is 90 degrees. The angle spanned by the hand spread out and held at arm's length is about 20 degrees. The

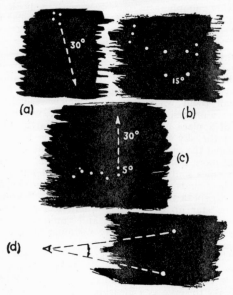

Figure 8

width of a finger at arm's length is about 2 degrees. An inch at a distance of a hundred yards represents an angle of one minute of arc.

We specify all the apparent separations between stars by the angle which the lines to them make at the eye. This angle in the case of the pointers of the Plough is about 5 degrees. More usually we say that the *angular separation* between these two stars is 5 degrees. Other sample separation angles are the following: the Great Square in Pegasus has a side of about 15 degrees. (Figure 8 (b)). The separation between the two stars, Castor and Pollux, which are the principal stars in the constellation Gemini, the twins, is about 5 degrees. The Plough is about 30 degrees from the north pole of the heavens (Figure 8 (c)) : the Southern Cross about the same distance from the South Pole of the heavens. (Figure 8 (a)).

The co-ordinates corresponding to latitude and longitude on the earth are measured on the sky in a very similar way. The angular distance of a star north or south of the celestial equator is called *declination*, is always denoted by the Greek letter delta (δ) and is measured from zero to 90°, positive values north of the celestial equator, negative values south of it.

Declination

Having got the idea of declination, let us see what we can do with it. First we must remember that the shapes of the constellations do not change perceptibly, so that the declination of a star is a number which is almost fixed. There are some obvious questions which we can answer when once we have the idea of the declination of a star at our disposal. For instance, we know that, watch we never so long, in the latitude of Britain we shall never see the Southern Cross, while from South Africa or Australia we shall never see Ursa Minor: which stars then can we see from which latitudes?

Imagine ourselves in the northern hemisphere at a moderate northerly latitude; we note the fact that the stars rise in the east, reach their greatest altitude above the horizon when they are due south, and then decline to set in the west. Clearly then, as far as being visible goes, a star is doing its best when it is due south, or as we say, on the *meridian*. The altitude of the pole is equal to the latitude of the observer. Ninety degrees southward from this brings us to a point on the celestial equator which is $(90 - \varphi)$ degrees above the horizon. As long as the star is less than $(90 - \varphi)$ south of the equator, i.e. if its south declination is less than $90° - \varphi$, it will be visible from a place in north latitude φ. It will also be clear that *some* stars will still be visible even when they are doing their worst, that is when they have swung round below the pole and are due north of the observer. These stars will be visible on any dark clear night, and are called *circumpolar* stars. A star will be circumpolar as seen from north latitude φ if it is not so far from the north celestial pole that it gets below the horizon even when it is due north and below the pole. Now the altitude of the pole is φ, so that a star will be circumpolar if it is north of declination 90° — φ. This is the same number as the one derived in solving the problem of which stars are visible at all. There is clearly a compensation: a certain area of the sky south of south declination 90° — φ is invisible, but to make up for that, an area of sky north of north declination 90° — φ is always visible. The stars which lie in declinations intermediate between these rise and set. (Figure 9.)

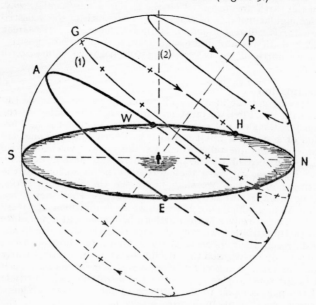

Figure 9

Star (1) rises at F, sets at H; Star (2) is circumpolar. P is the North Celestial Pole; EAW the Celestial Equator.

We could put exactly the same argument for a place south of the equator. For such a place the stars near the north celestial pole (north of 90° — *south* latitude of observer) are invisible, but to make up for this, the stars south of declination (90° — south latitude of observer) are always visible and are circumpolar round the south celestial pole. We must note that being circumpolar or being visible is something which depends not only on the declination of the star but also on the latitude of observation. Let us apply our rules for example in the case of a man who lives at the north pole of the earth.

Now $\varphi = 90°$. Hence all stars south of the celestial equator are invisible, since $90° - \varphi = 0$. On the other hand, any star north of the equator is circumpolar. As the earth rotates the stars go around parallel to the horizon, and no star rises or sets. In fact, as we can see in several ways, every star which is visible is circumpolar, and the intermediate zone of the sky comprising the stars which rise and set has shrunk to nothing. The same is true at the south pole of the earth. All the stars south of the celestial equator are circumpolar; all the northern stars are invisible.

For a man who lives on the equator of the earth the situation is entirely different. Now the value of φ is zero, so that, since declinations run only up to $90°$, there are no circumpolar stars, and all the stars rise and set. From the equator of the earth it is possible to see every star in the sky at some time or other, but none of the stars remains above the horizon all the time. If you like you can say that from the poles of the earth you can see half the sky all the time: from the equator you can see all the sky for half the time. At the poles the stars on the horizon just skim around it as the earth rotates. A little further south those stars which rise and set do so at a very small angle to the horizon. On the equator the stars go straight down and set at right angles to the horizon.

The Midnight Sun and the Polar Night

We shall deal in detail with the sun later. For the present we may regard the sun as a star which moves against the background of the other stars and whose declination varies from $23\frac{1}{2}$ degrees north in the northern summer, to $23\frac{1}{2}$ degrees south in the northern winter. By what has already been said, there must clearly be some parts of the earth for which the sun is a circumpolar star. When it has a declination of $23\frac{1}{2}$ degrees north, it will be circumpolar north of latitude $66\frac{1}{2}$ degrees. This is the latitude of the Arctic Circle: when the sun is at its farthest north it is circumpolar at all places north of this latitude, that is, it is above the horizon throughout the 24 hours and thus we have the phenomenon of the midnight sun. But at the same time as the midnight sun is occurring in the north, the polar night reigns in the south, south of the Antarctic circle at $66\frac{1}{2}$ degrees south latitude, for our formula shows us that south of this latitude the sun is a star which is never visible. In the northern winter, when the sun has gone to south declination 23½ degrees, conditions are reversed: the north has the polar night, the antarctic regions the midnight sun. On two intermediate days of the year, in spring and autumn, the sun has zero declination and lies on the celestial equator. Then whatever the latitude of the observer, the sun rises exactly in the east and sets exactly in the west, remaining above the horizon for exactly half of the 24 hours.

The consideration of the angle of rising and setting applies equally well in the case of the sun. If the sun is visible at all, then the further north we go the smaller is the angle to the horizon which the path of the rising or setting sun makes. In the far north it takes a perceptible time for the sun to pass below the horizon, a time still further lengthened by certain effects of the earth's atmosphere which we shall consider in Chapter V. Twilight may therefore last for a long time. On the other hand, near the equator, the sun, like the stars, rises and sets perpendicular to the horizon, and thus occupies the smallest possible time in the process. Twilight in the tropics therefore lasts only a few minutes. The well-known abruptness of the rising and setting of the sun in the tropics is thus due to the angle to the horizon at which it rises and sets.

The Rotation of the Earth

We have hung this argument on a navigational peg, introducing it as the method of finding the substellar points of stars. So far we have dealt only with declination, and, from what has gone before, it should be clear that the substellar point of a star in north declination δ is in north latitude δ— somewhere on this parallel of latitude the star is in the zenith, but the exact location, east and west, of the substellar point changes as the earth rotates.

To proceed further we must consider the question of the rotation of the earth or, what is the same thing, the appearance of the sky as it rotates past the observer. One fundamental line on the observer's celestial sphere has already been mentioned—the meridian—the line which, in the northern hemisphere, extends south from the north celestial pole, through the zenith, down to the south point of the horizon. In the southern hemisphere we regard the meridian as extending from the south celestial pole, through the zenith and up to the north point of the horizon. As we have already seen, a star will have its greatest altitude when it crosses the meridian. For each and every observer, the sky looks like a sphere with the star positions marked inside it, which rotates as a rigid body, the axis of rotation being parallel to that of the earth.

This axis is, therefore, tilted up at an angle corresponding to the latitude of the observer.

Think of a peeled orange with the lines of the " divisions " all running from one point at the bottom of the orange to a point at the top. We can think of the celestial sphere as being marked out in this way, the top and bottom corresponding to the celestial poles, while the celestial equator is the extra line which would be made if we sliced the orange in halves the way grapefruit are cut.

If we mark a point on one particular " division " on the orange, and turn it on a knitting needle stabbed through the fruit from pole to pole, then we can always specify the position of the orange by saying how far round we have turned the selected division line. We measure the angle through which the orange has turned at one or other of the poles—the value will be the same—nor does it matter where on our selected division we have made our mark, the value will always be the same.

If then in imagination we draw a line up from any star to the north pole of the heavens, this line will make a certain angle at the north or south poles with the meridian of the observer. As the earth rotates from west to east, this angle, starting from zero when the star is on the observer's meridian, will increase steadily as the star moves west. The important point is that the angle increases uniformly—just as the angle between the small hand of a clock and the 12 o'clock position increases uniformly—and at approximately the rate which would apply to a clock graduated for twenty-four hours— one twenty-fourth of 360°, or 15 degrees, every hour. Thus a star on the celestial equator (zero declination) which is now on the meridian will have moved a quarter of the way round the sky after six hours, and, since the equator and the horizon intersect at the east and west points, it will, after this lapse of time, be just setting due west on the horizon. The movement of the circumpolar stars is more readily seen: if you watch for only half an hour you will be able to see a perceptible shift in the position of either the Plough or the Southern Cross.

The angle formed at the north or south celestial pole by the meridian and the line to any star is called the *hour angle* (H.A.) of that star and it increases at approximately 15 degrees per hour. (Figure 10.) This idea is particularly important because as may be clear already, the hour angle of a star does not alter if the observer is transported instantaneously on the surface of the earth due north or south. It depends on the longitude of the observer and the particular star observed, but not on the latitude of the observer. That it depends on the longitude of the observer is easily seen, for clearly, places further east will meet a star being carried from east to west by the rotation of the sky, sooner; that is, points further east on the surface of the earth will, at the same moment of time correspond to larger hour angles for any selected star.

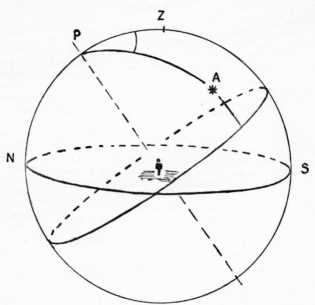

Figure 10

Z PA is the Hour Angle of Star A.

In order to set up a workable system of star positions, we must have a starting point, or origin, just as we do on the surface of the earth. On the earth the origin of our system of latitude and longitude is the point in latitude zero (on the equator) and in longitude zero (on the meridian of Greenwich). In the same way we have a fundamental point on the sky which marks the origin of our system of map references for stars. It is a point on the celestial equator which is variously called the *First Point of Aries*, or the *Vernal Equinox*. The way in which it is defined and the reason for giving it these names will appear later. For the moment it is just an arbitrary point on the sky from which we start our system of reckoning. We have already defined and discussed declination which corresponds to latitude on the earth. Now we

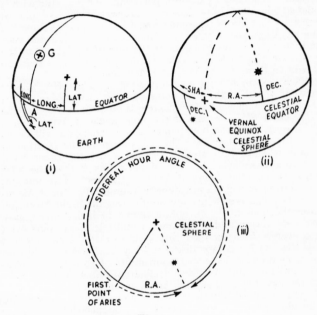

Figure 11

(i) Defining positions on the Earth; (ii) positions on the Celestial Sphere; (iii) Relation between R.A. and S.H.A.

must discuss the system of reckoning corresponding to longitude. (Figure 11.)

Right Ascension and Sidereal Hour Angle

There are two methods of doing this: one the method of the professional astronomer, the other a variant introduced by the needs of astronavigation for aircraft during the war. The second is perhaps the simplest to understand. In this system we reckon the position of a star by its " longitude " round the celestial equator reckoned westwards from the first point of Aries in degrees. This angle is called the Sidereal Hour angle (S.H.A.) of the star and runs from 0° to 360°.

The astronomers reckon in the reverse direction from the first point of Aries and do so in terms of hours, minutes and seconds of time, reckoning 15 degrees per hour. Thus a star with a sidereal hour angle of 60 degrees is 60° west of the first point of Aries. If one starts reckoning round in the opposite direction one has to go 300 degrees (360° minus 60°) to get to the star, and, converting this to hours at the rate of 15 degrees to the hour, one gets 20 hours. The astronomer then says that the star has a *Right Ascension* (R.A.) of 20 hours.

Each system has its own advantages. The SHA tells us how far ahead of the first point of Aries any star is. Thus the star with an SHA of 60 degrees is 60 degrees ahead of the First Point of Aries, and if the hour angle of Aries is, say, 40 degrees, then the hour angle of the given star will be 60 plus 40 = 100 degrees. The calculation of the position of a star is thus very straightforward. On the other hand the same calculation in professional terms is that the hour angle of the First Point of Aries is 2 hours and 40 minutes, (converting 40 degrees into time measure). The Right Ascension of the star is 20 hours, and subtracting the Right Ascension from the hour angle we obtain minus 17 hours 20 minutes. We may always add or subtract a multiple of 24 hours or of 360 degrees from such a result so as to make it positive and less than 24 hours (360 degrees). The reason for this should be clear if one reflects that hour angles of 30 hours or minus 18 hours for a star of a given declination define the same point in the sky for that star as an hour angle of 6 hours: this value is obtained by subtracting 24 hours from the first result, and adding 24 hours to the second, an example which illustrates the quoted rule.

Applying this in the present case (adding 24 hours) we arrive at an hour angle of 6 hours 40 minutes, which is the same as 100 degrees converted into time measure at the rate of 15 degrees per hour. We can write the general result as,

Hour Angle (H.A.) of star = H.A. Aries + S.H.A. star

or = H.A. Aries − R.A. star.

Sidereal Time

At any place, the hour angle of the First Point of Aries is called the *local sidereal time*, and the hour angle of the First

Point of Aries at Greenwich is called the *Greenwich Sidereal Time*. Greenwich Sidereal Time is not the time shown by an ordinary clock, but in books like the *Nautical Almanac* and the *Air Almanac*, the Greenwich Sidereal Time is tabulated for any moment of ordinary time. The particular advantage of the Right Ascension system is that each object comes on to the meridian of a place when the local sidereal time there is equal to the R.A. of the object.

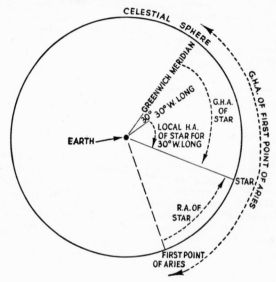

Figure 12

Now we are in a position to see how all this apparatus is used to fix the position in the sky of any star at any time. First let us suppose that one is at Greenwich. At the given time one looks up the Greenwich Hour Angle of the First Point of Aries, that is the Greenwich Sidereal Time. Then one looks up in the table of star positions the R.A. or SHA of the selected star. This, combined with the given value of the Greenwich Sidereal Time gives the Greenwich Hour Angle of the selected star. We can then draw a diagram or make a model of the celestial sphere on which we can mark in exactly where the selected star will appear at the given time as seen from Greenwich. But, in navigation, one wants to move about the globe and to have methods of knowing the position of a star as seen from any place at any time. How then is the local hour angle of a given star related to the Greenwich Hour Angle of the same star at the same moment?

We can see this best by drawing a picture showing the north pole of the earth. (Figure 12.) Clearly if one is in a certain longitude, say, 30 degrees west of Greenwich the hour angle of any star will be 30 degrees less than it is at the same moment at Greenwich. We can therefore write down the simple formula for the local hour angle of any star as seen at any time from any longitude:

Local hour angle of star = Greenwich hour angle (G.H.A.
of star — west longitude.

or, by what has already been said

Local hour angle of star = Greenwich hour angle of First
Point of Aries +

SHA of star — longitude of
observer

or, Local H.A. Star = G.H.A. Aries — R.A. Star — W. Long. of observer. In doing such calculations east longitudes count negative and must be added. The answer must also lie between 0° and 360° (0^h and 24^h) and if it does not, one adds or subtracts multiples of 360° (24^h) to make it do so.

We now give some examples of this type of calculation. The two given contain most of the essentials of this type of problem. When you have run through the whole of this chapter come back and study them. Note that, although stars always go westwards, this direction is *into* the page in Figure 15 (a northern celestial sphere with the north celestial pole at the top), and *out of* the page in Figure 16 (a southern celestial sphere with the south pole at the top). Only practice can give mastery of problems of this kind and you should try propounding and solving others for yourself using the data given in the appendices.

Examples

(1) The declinations of Spica and Vega are respectively 10° 54′ south and 38° 45′ north. From what latitudes are they visible as stars which rise and set and in what latitudes are they circumpolar?

Imagine an observer to be in north latitude φ. Then when Spica is on the meridian, due south, it will be 10° 54′ below

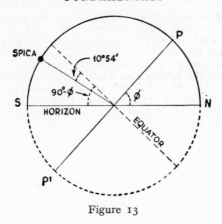

Figure 13

the south point of the celestial equator. (Figure 13.) The altitude of the latter point is $90° - \varphi$, so that at this moment the altitude of Spica is $90° - \varphi - 10° 54' = 79° 6' - \varphi$. If this is positive Spica will be above the horizon i.e. if φ is less

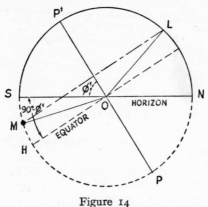

Figure 14

than $79° 6'$. Spica will thus be visible either as a star which rises and sets or as a circumpolar star at all latitudes south of $79° 6'$ north. To find in what latitudes Spica will be circumpolar we draw a celestial sphere (Figure 14) for an observer

B

in a southerly latitude φ′. Spica travels on the line LM and is circumpolar if M is above the horizon. This is the case if angle MOH (= 10° 54′) is greater than 90° — φ′ i.e. if φ′ is greater than 90° — 10° 54′ = 79° 6′, Spica is thus circumpolar for observers south of south latitude 79° 6′.

For Vega the results are: Vega is visible north of south latitude 90° — 38° 45′ = 51° 15′ and is circumpolar north of north latitude 51° 15′. In most parts of England Vega is just above the horizon when due north underneath the pole as it is on winter nights.

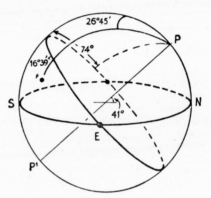

Figure 15

(2) If the Greenwich Hour Angle of the Vernal Equinox is 3h draw diagrams to show the positions of (i) Sirius as seen from Istanbul and (ii) the Southern Cross as seen from Sydney at this moment.

Sirius: R.A. 6h 43m: declination — 16° 39′ and

α Crucis: R.A. 12h 24m: declination — 62° 49′

Istanbul: Longitude 29°E Latitude 41°N

Sydney: Longitude 151°E Latitude 34°S

(i) Draw a celestial sphere with the north pole at an altitude of 41° corresponding to the latitude of Istanbul. (Figure 15.) The G.H.A. of the Vernal Equinox is 3h, corresponding to

45°. Istanbul is 29° east of Greenwich. Therefore the Istanbul hour angle of the Vernal Equinox is

$$29° + 45° = 74°.$$

The Vernal Equinox is thus 74° *west* of the meridian.

Sirius is 6h 43m east of the Vernal Equinox. In Angular measure

$$6h = 90° \text{ and } 43m = 10° 45'$$

i.e. Sirius is 100° 45' east of the Vernal Equinox or

$$100° 45' - 74° = 26° 45' \text{ } east \text{ of the meridian.}$$

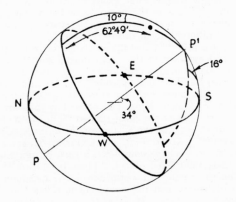

Figure 16

Mark this angle on the diagram and put in Sirius 16° 39' south of the equator. The angle 26° 45' is 1h 47m of time, i.e. in Istanbul Sirius is east of the meridian by 1h 47m and is about 30° above the horizon.

(ii) Draw a celestial sphere with the south celestial pole at an altitude of 34° corresponding to the south latitude of Sydney. (Figure 16.) The Greenwich hour angle of the Vernal Equinox is 45° and Sydney is 151° east of Greenwich. Therefore the Sydney Hour Angle of the Vernal Equinox is 151° + 45° = 196° = 180° + 16°. α Crucis (the brightest star in the Southern Cross) is 12h 24m *east* of the Vernal Equinox or in angular measure 186° east, i.e. α Crucis is 186° − 196° = − 10° east of the meridian. The Southern Cross is thus

+ 10° *west* of the meridian. Mark this angle at the pole and put in the Southern Cross 62° 49' south of the equator. The Southern Cross is therefore almost due South and at an altitude of about 60°.

Sidereal Time and Solar Time

The Greenwich Hour Angle of the First Point of Aries is tabulated in the *Nautical Almanac* or the *Air Almanac*. These tables are calculated in advance, and a sketch of the methods used is of interest to us. This will make clear the usage, which may seem somewhat odd, of calling the Greenwich Hour Angle of the First Point of Aries the Greenwich Sidereal Time. What is the meaning of the phrase Sidereal Time? Sidereal means pertaining to stars : why should star time be different from ordinary time?

The stars are infinitely distant as compared with terrestrial distances, and even when the earth moves round the sun to occupy, after six months, a position in space which has changed by nearly two hundred million miles, the apparent positions of the stars will change only by amounts far too small to be considered here. Such changes do occur but they are detectable only by the use of the most refined technical methods.

For all purposes then, the system of the stars forms a fixed reference system. As we rotate on our terrestrial roundabout, we can tell when we have completed a turn by noticing when we pass a particular fixed object. Thus, by noting the interval between the passage of a star across our meridian on two successive nights we can tell how long it takes the earth to rotate once on its axis. If we perform this experiment we shall find that there is an interval of 23 hours and 56 minutes by the time shown on an ordinary clock between the meridian passages of a star on two successive nights. This time interval of 23 hours 56 minutes of ordinary time is called a sidereal day, and is the true time occupied by the earth in making one revolution on its axis. The sidereal day is divided into 24 sidereal hours, each of these into 60 minutes of sidereal time, and each of these again into 60 seconds of sidereal time. In one sidereal hour the earth turns one twenty-fourth part of a revolution, or through 15 degrees. Earlier on, we spoke of the earth turning about 15 degrees per hour. Now we can make this statement precise. It turns through 15 degrees, and the hour angles of all stars increase by 15 degrees, in one *sidereal* hour. The use of the approximation arose because of the difference between ordinary time and sidereal time.

The reason for this difference is that the earth is not only rotating on its axis, but is also moving in an orbit round the sun, taking about 365¼ days to make a complete circuit. To understand the effect of this, make a model. Put a lamp on the floor and walk round it sideways facing directly towards the lamp. The lamp, representing the sun, remains always on the meridian (represented by the direction of your nose). The fixed system of stars is represented by the objects in the room. When the circuit is complete you will find that although the " sun " has not moved in your sky, you have faced every part of the room in turn. The sun has been in front of every part of the room in turn. You have made one turn as judged

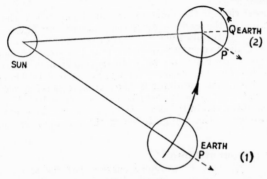

Figure 17

by the nxed objects in the room. Now, if you can, waltz round the lamp (reverse turns all the time) making 365 turns past the lamp—that is, you must make the lamp sweep past your nose 365 times. If after this performance you are still capable of counting, you will discover that you have still made one extra turn as judged by the fixed objects in the room. There is always this difference of one turn for each complete circuit in the case of a body rotating on its axis and simultaneously describing a circuit round a fixed centre.

This means that 365¼ turns as judged by the movement relative to the sun will correspond to 366¼ turns as judged by the fixed stars. A turn relative to the stars will take less time than a turn judged by the sun, the defect being 1/365 of the time required for one turn. One day divided by 365 is just less than four minutes of time, which is just the difference between a sidereal day and an ordinary day. In the course of a year, all the four minute periods will just add up to one day.

We can look at the matter from a different point of view. To imitate the sun you must make reverse turns in your waltz so that you are turning on your axis in the same direction as your turning about the central sun. If you or the earth turn enough to bring a star back on to the meridian you will find that, since you have also advanced a little round a curved orbit, you need to turn still a little further in order to bring the sun back on to the meridian. This extra turn is through an angle of about 1 degree, and since the earth turns 15 degrees in the hour, this occupies a time of about four minutes. (Figure 17.)

Sidereal time is the more fundamental idea, since the rate of turning of the earth against the stars is very closely uniform. But naturally since a sidereal clock gains on an ordinary clock, the two brands of time are not in step. A sidereal clock gains 12 hours on an ordinary clock in six months, and is therefore useless for ordinary day to day timekeeping. We must regulate our affairs by the sun, and therefore we need a different system of time reckoning which is based on the position of the sun in the sky. We must get up in the morning round about sunrise and go to bed in the evening round about sunset.

The Equation of Time

It is fairly easy to convince oneself that the sun itself is not a particularly good standard timekeeper. As we know, the sun will be at its greatest altitude when it is on the meridian, so that, by putting a piece of stick in the ground and noting when the shadow is shortest, we can find the moment at which the sun crosses the meridian. Or we can set up two sticks or a sheet of board to mark true north and south (not magnetic north, which differs from true north in most parts of the world), and note the time when the sun crosses the line. If one notes the time by an ordinary clock at which the sun crosses the meridian one will soon find that there is a difference between what is called apparent noon (the moment when the sun is on the meridian) and noon as shown by a clock, which is variable according to the season of the year and which may amount to nearly twenty minutes of time at its greatest. The energetic seeker after astronomical knowledge can do this experiment, say, every Sunday for a year and make a graph of the results. The armchair scientist can convince himself of the result in an easier way. As we have already seen, the sun rises in the east and sets in the west due to the rotation of the earth, and half-way between rising and setting is on the meridian at its greatest altitude.

Look up the times of sunrise and sunset given in your pocket diary, and for each day find out the time half-way between the two. Thus if the sun rises at 6.10 a.m. G.M.T. (that is 5 hours 50 minutes before noon) and sets at, say, 6 hours 5 minutes after noon, then half-way between the two times is 7m 30s after noon, which will represent roughly (but only roughly) the difference between sun time (what is called apparent solar time) and Greenwich Mean Time. This difference is called the *equation of time* from an old, and, nowadays, almost paradoxical, use of the word equation, to mean something which has to be added to make an equality.

Now let us consider how this difference comes about, and why the sun itself is a bad time standard. In constructing our model we remarked that as one moved round the central lamp, it came successively in front of various objects in the room. In the same way, as the earth moves round the sun, the sun is seen, or would be seen if it were not so bright, against the background of different constellations during different seasons of the year. We can therefore think of the sun as a very bright star which moves against the background of the other stars making a complete circuit of the sky in the course of a year. The apparent path of the sun against the background of the stars is called the *ecliptic*.

Now the path of the earth around the sun is not circular, but oval in shape, and the rate at which the earth moves in its orbit varies, being more rapid when the earth is near the sun (which it is during winter in the northern hemisphere) than when the earth is more remote from the sun. This fact is a consequence of the law of gravitation which governs the motion of the earth in its orbit. Thus the motion of the sun against the background of the stars is not uniform. As we have seen, the difference between the length of the ordinary day and the sidereal day is due to the fact that each day we have to wait a little after the earth has completed a revolution as judged by the stars to catch up with the sun. If the sun were losing on the stars at a uniform rate, this difference would be constant. But in fact the sun is losing at a rate which is now greater than the average and now less. If we replaced the actual sun by a fictitious body moving round the ecliptic in a year at a uniform rate, then the real sun would sometimes be ahead of the fictitious body, and sometimes behind.

There is a second cause of the difference between solar time and Greenwich time. The ecliptic is a path which is inclined to the equator. This is a consequence of the fact that the axis of the earth is tilted. Instead of being perpendicular to the plane of the orbit of the earth around the sun the earth's axis is tilted at an angle of about 23½ degrees to this perpendicular. The axis points always to the same point in the heavens—the north celestial pole—so that at one time in the year the north end of the earth's axis is tilted in towards the sun, while six months later it is tilted away. (Figure 18.)

If, therefore, we were to mark out against the stars the track followed by the sun in the course of a year, we should find that it was a circle round the sky tilted with respect to the celestial equator. The celestial equator forms the basis of the standard system of reference on the sky, so that the points on this circle

Figure 18

followed by the sun (the ecliptic) are at different declinations. As the sun moves round the ecliptic its declination changes gradually from + 23½ degrees to − 23½ degrees, and back again. We have already used this fact in the explanations of the midnight sun. Now we see why it is that the sun's declination is variable, being 23½°N in the middle of the northern summer, when the northern end of the earth's axis is the one which is inclined towards the sun, and 23½°S in the middle of the southern summer (northern winter), when it is the southern end of the axis of the earth which is nearer the sun.

The phenomena of the seasons, the midnight sun, and the polar night are the result of this tilting of the earth's axis. Although the distance of the earth from the sun, its source of light and heat, does vary in the course of a year because the orbit of the earth is oval and not circular, the proportionate variation is small. The varying distance has little to do with the variation of temperature between summer and winter, this being almost wholly due to the variation in the apparent height of the sun above the horizon, and the varying number of daylight hours out of the twenty-four.

The Mean Sun and Greenwich Mean Time

The difference between solar time and mean time (the equation of time) is due partly to the non-uniform motion of the sun in the ecliptic. It is also partly due to the inclination of the ecliptic. If the sun were moving uniformly along the ecliptic, then at the points where its path crosses the celestial equator (the equinoxes) where it is moving at an angle to the

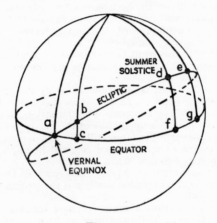

Figure 19

ecliptic, the rate of increase of the sun's right ascension would be less than the rate of movement of the sun in its path. (Figure 19.) Part of the rate is, as it were, made ineffective because the sun is moving partly sideways to the equator. But when the sun is near the most northerly and southerly parts of its path it is moving parallel to the equator and the full value of the daily increase of right ascension will be effective. It will be clear that the time when the sun crosses the meridian each day depends on its right ascension (or sidereal hour angle) so that an effective time-keeping body would be one whose right ascension increased at a perfectly uniform rate throughout the year. For the two reasons outlined the real sun does not do

B*

this. For time-keeping purposes the real sun is therefore replaced by the *mean sun*, a fictitious body which goes round the equator at a uniform rate adjusted so that the real sun and the mean sun each take a year to make a circuit. Greenwich Mean Time is the Greenwich Hour Angle of the mean sun plus 12 hours. The 12 hours is added so as to make the day start at midnight.

We have incidentally cleared up several outstanding points in the course of this argument. The vernal equinox (or the First Point of Aries) is one of the points of intersection of the celestial equator with the ecliptic. When the sun is at this point it is on the equator; it therefore rises due east and sets due west, and is above the horizon exactly 12 hours. Day and night are equal; this day occurs in spring about March 22 or 23 and the explanation of the name " vernal equinox " (pertaining to spring; equal night) is clear. Right ascension is measured from this point in the direction such that the right ascension of the sun is always increasing in its movement against the star background.

Three months later, about June 22 or 23 the sun is at its most northerly position, with a right ascension of 6 hours and a declination of $23\frac{1}{2}$ degrees. This situation is called the summer solstice, and is the middle of the northern hemisphere summer. Three months later still the sun is crossing back over the equator to the southern celestial hemisphere. This is the autumnal equinox, when again, the sun being on the equator, day and night are equal. Lastly about December 22 or 23 the sun is at its furthest south and the situation is called the winter solstice.

Time-keeping is thus a more complex business than one might have thought, and even now we have merely referred very simply to the most important elements in the situation. To summarise, these are: uniform time is defined by the rotation of the earth, giving sidereal time. This being unsuitable for civil requirements, a time reckoning system based on the sun is required. The sun itself is unsuitable because, by reason of its non-uniform motion against the stars, and the inclination of its path, the moments of the passage of the sun across the meridian of any place are not equally spaced. The sun is therefore replaced by a fictitious body, the mean sun, which crosses the meridian at uniform intervals. The Greenwich Hour Angle of the Mean Sun (in time measure) plus 12 hours, is Greenwich Mean Time. It differs from the time based on the true sun by an amount which varies throughout the year which is called the equation of time. (Figure 20.)

Local Times and Zone Times

This defines time at Greenwich but not for any other meridian on the surface of the earth. In fact, of course, Greenwich time is kept all over the British Isles which means that the apparent time as shown by the position of the sun will be affected by the longitude for places east or west of Greenwich. But it would be most inconvenient if each town kept local mean time so the plan of keeping a standard time throughout

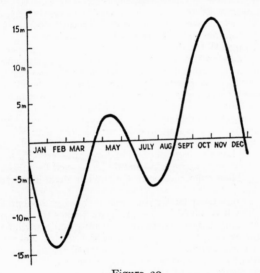

Figure 20

The Equation of Time. The graph shows the value of Apparent Time – Mean Time throughout the year.

a zone of longitude is adopted. In the British Isles this zone time is Greenwich Mean Time. In the west of England and Wales, times of phenomena, such as sunrise and sunset, will all be late by a few minutes due to the change of longitude to the west. This will affect the armchair method of finding the equation of time mentioned above and for a given place will shift the whole graph of Figure 20 up or down by a few minutes. To get the true values you must apply a correction corresponding to your longitude difference from Greenwich converted into time.

Sundials and suncompasses (some forms of the latter) require the equation of time and the longitude difference to be taken into account.

In other countries time zones corresponding to a selected meridian passing through the country are adopted. The Atlantic Coast of the U.S.A. keeps the time of the meridian 75 degrees Longitude or 5 hours back from Greenwich. The Union of South Africa keeps the time of the meridian 30 degrees E or 2 hours fast on Greenwich—and so on all over the world. Approximately at 180 degrees longitude is the international date line. Places west of Greenwich keep time slow on Greenwich, the difference increasing the further west one goes, until at 180 degrees W the difference is 12 hours. Going east, zone times get more and more fast on Greenwich until at 180 degrees E longitude the zone time is 12 hours fast. 180 degrees E and 180 degrees W represent of course the same meridian and by convention in crossing this line one jumps a whole day, gaining a day in one direction and losing it in the other.

In dealing with problems in which zone time is involved the simplest procedure is first to convert the zone time to Greenwich. For example a problem concerned with an observation made in W Longitude 73 degrees at local time 10 a.m. (corresponding to zone 75 degrees W) should be tackled by considering it as a problem concerned with Greenwich Mean Time 10h + 5h = 15h.

We may now come back to our original navigational problem, and see how it is solved. The navigator wishes to know where the substellar point of a particular star may be. It will be now clear that he can, from the known G.M.T. of his observation, find the corresponding Greenwich Sidereal Time. The substellar point is in a latitude (north or south) equal to the north or south declination of the star. Its west longitude will be equal to the Greenwich Hour Angle of the star which he can find from the R.A. of the star, and the Greenwich Sidereal Time for the moment concerned. By methods which cannot be discussed here, he can then work out the position circle corresponding to the observed altitude, and, repeating this for a second star, he is enabled to fix his position as the intersection of two position circles.

The Calendar

We conclude this chapter with a few brief remarks on the subject of calendars. We, in these days, are apt to think of a calendar as something bearing a picture of a pretty girl, which the grocer sends round to his customers at Christmas, recording a scheme of dates which are, somehow, fixed from time immemorial. We fail to realise that calendars are invented and that our our present calendar is an ingenious compromise between a variety of almost irreconcilable astronomical facts, mixed up with a great deal of history and tradition.

Let us ask ourselves why we need a calendar, what conditions must it satisfy, and how we would set about re-establishing a calendar on the basis of our own observations. Calendars are, fundamentally, based on the two facts that there are certain obvious astronomical periodicities, and that for most purposes it is convenient to lump days into a series of blocks (such as months or years) rather than to adopt the course of numbering the days in order, starting from some arbitrary date.

The most obvious astronomical periodicity is the recurrence of the phases of the moon, a period which is somewhat irregular, but which averages 29·5306 days. Most early calendars were based on this period, or on the nearest round number of 30 days, and traces of this practice remain in our own months which are all close to 30 days in length.

The average interval between successive passages of the sun through the vernal equinox (what is called the tropical year) is 365·2422 days. The recurrence of the seasons takes place in this period. At this interval the sun reaches its greatest northerly declination. The time between sunrise and sunset is then longest. (See Figure 9.) Spaced at the same intervals, are the moments when the sun has its greatest southerly declination, when the daylight hours in the northern hemisphere are shortest. While it is true that the recurrence of seasonal weather conditions is far less pronounced and less easily recognisable than the recurrence of lunar phases, it is important for agriculture to know in advance when (on the average) spring or autumn weather may be expected. This is most easily secured by fixing a definite date in advance. Now, if the length of the year is wrongly estimated by the calendar in use, any error will accumulate steadily, so that, in the end, such a calendar will lead to seriously erroneous predictions of the advent of seasonal weather conditions. As an example, take the case of the Mohammedan Calendar which is entirely lunar, having alternate months of 29 and 30 days, giving a total year length of 354 days. In any given year the error of 11¼ days may not be serious, but in two years this will have become 22½ days, and so on, so that the lunar calendar very rapidly falls out of step with the solar (or seasonal) calendar. This calendar is still in use, and, for example, the fast month, Ramadan, circulates round the year quite rapidly.

It is therefore essential for modern purposes to arrive at a compromise, which preserves the useful division into months, and which yet gives a close approximation to the true length of the year. Attempts to secure this condition have been very numerous, and we cannot go into details. An important one was the calendar adopted by Julius Caesar on the advice of Sosigenes, the Alexandrian astronomer. Julius Caesar added 10 days to the old Roman lunar year of 355 days and introduced a leap year (in which February had 29 days) every fourth year. After many vicissitudes and misinterpretations the calendar was stabilised in A.D. 325 so as to follow this rule, which gives a year length of 365·25 days. In this period, the beginning of the year was fixed as March 25th, to coincide with the Feast of the Annunciation and, approximately, with the Vernal Equinox. Thus, in this system, now known as " Old Style ", the dates, March 1st 1000 A.D. and April 1st 1001 A.D. were separated by only one month.

This, the Julian calendar, gives a year length of 365·25 days, whereas the required true length is 365·2422 days. The annual difference of 0·0078 days will, in a century, accumulate to more than three-quarters of a day, or exactly, one day in 128 years. After this calendar had been in use for a thousand years, the error had accumulated to serious proportions, and, in 1582, Pope Gregory reformed the calendar so as to eliminate this error. This was done by omitting several days in the reckoning so that the day following October 4th 1582 was October 15th 1582. This jump was slightly larger than necessary, and was made on astronomical advice, in order to adjust the date of the vernal equinox. To prevent a recurrence of the error, the device was adopted of dropping leap years in century years, except when the first two figures were also divisible by four. This system, the Gregorian calendar, is that which we use today : the year 1900 was not a leap year, but 2000 A.D. will be. The result is to reduce the length of a century by 0·75 days, instead of the 0·78 required, but the residual error is so small that it will not accumulate to serious proportions for many centuries.

England adopted the Gregorian calendar in 1752 by which time the error had accumulated to 11 days. September 2nd 1752 was followed by September 14th 1752, and at the same time the date of the beginning of the year was changed to January 1st. At the time, riots, with the cry " Give us back our eleven days " broke out. The principal modern relic of this situation is provided by the fact that the British financial year ends on April 5th, which is Old New Year's Day (March 25th) displaced by the eleven days change of 1752.

There is some opinion in favour of further calendar reform so as to bring the day of a given date on to a fixed day of the week. This would render it unnecessary to print new calendars for each year, and would be achieved by the use of a Year Day every year, and a Leap Day every fourth year, neither of which would be given the name of a weekday.

CHAPTER II

WHAT'S IN THE SKY?

We now have a system of map references (Right Ascension and declination) for defining the position of any star: given these numbers we can estimate roughly the position in the sky of the star to which they correspond as seen at any time from any part of the earth. But so far our sky is like a new sheet of graph paper or a blank map. It is ruled with lines of declination and right ascension but it has no stars on it.

Star Magnitudes

Astronomers use a system, called the magnitude system, to describe the brightness of various stars as they appear to us. The system owes its origin to two facts, one historical, one physiological. In ancient times the astronomer Hipparchus, wishing to describe the brightness of stars, arranged them in order. In this hierarchy of splendour the brightest stars were said to be of the first magnitude. Then came a group of prominent, but less bright stars, described as being of the second magnitude. Third magnitude stars are readily visible to the unaided eye, but are usually the less prominent ones of each constellation. Fourth and fifth magnitude stars are fainter still, and sixth magnitude stars are those which are just visible to the naked eye. There are, of course, fainter stars which can be seen with the aid of telescopes. The total number of stars visible to the naked eye in the whole sky is somewhat less than five thousand, a surprisingly small number.

In this way of specifying the brightness of a star the larger magnitude numbers correspond to fainter stars. It is not a very precise system, because it groups together in each class, stars having a range of brightness. In fact, the human eye is quite capable, with training, of distinguishing ten steps of brightness between each of the magnitude classes. In the nineteenth century it became necessary to replace the crude system of Hipparchus by something a good deal more precise. When precise measurement became possible it was found that the sixth magnitude stars were almost exactly one hundred times as faint as the first magnitude stars, and that each step of one magnitude corresponded to a constant ratio of brightness.

What this comes to is this: if the eye is presented with two series of lamps, of candlepowers, one, two, three, four, etc., all at the same distance, and a second series, of candlepowers, one, two, four, eight, etc., then it is the second series, not the first, which will appear to form a uniform graduation of brightness. In the first series each lamp is brighter than the next by a constant difference. In the second, each lamp is brighter than the previous one in a constant ratio (two). This property of the eye makes it possible for us to appreciate very great ranges of illumination at one glance. There is, in addition, an automatic adaptation of the eye to the general level of illumination which prevails, so that, for example, bright moonlight seems as bright as day, even though the intensity is several hundred thousand times smaller than that of sunlight. But within each picture, whether by day, by moonlight, or by starlight, we can appreciate a tremendous range of brightness. Just suppose for example that we count one star as being of brightness 10 and another one as being of brightness 20. Then since our eyes respond to equal ratios of brightness, ten steps of this kind (doublings) will carry us through the range represented by 10, 20, 40, 80, and so on up to 10,240. If on the other hand we saw the brightnesses 10, 20, 30, 40, etc., as equal steps, ten such steps would take us only to a brightness of 110. We thus appreciate a tremendous brightness range and, in addition, we find it just as easy to appreciate a ten per cent variation of brightness in a faint star as in a bright one. If our eyes responded to equal absolute changes, a ten per cent variation in a bright light would seem tremendous, whereas a ten per cent change in a faint light would not be detectable.

Thus when Hipparchus's system came to be refined, and when it was found that a sixth magnitude star was one hundred times as faint as a first magnitude star the intermediate steps to be inserted had to be such that each change of one magnitude corresponded to the same brightness ratio. The value chosen for this ratio was 2·512 because 2·512 multiplied by itself four times i.e. $(2·512)^5$ equals one hundred.

To fix the system completely it is only necessary to define the magnitude of one standard star, or better still to found the system on a group of stars of accurately-known brightnesses. The bright star Vega has a magnitude near zero. If it were exactly zero then a star with a magnitude 1·0 (notice that the system allows us to specify intermediate brightness by decimals of magnitudes) would be one whose brightness was $1/2·512 = 0·4$ times that of Vega. A star of magnitude

2·0 would be of brightness $(1/2·512)^2$ or 0·16 times that of Vega. The exact mathematical rule is that, L, the luminosity of a star, is related to the magnitude number m assigned to it, by the formula

$$L \text{ is proportional to } 10^{-0.4m}$$
$$\text{or } m = \text{constant} - 2·5 \log L.$$

Incidentally, if you are used to logarithms, don't forget that the usual practice in writing down the logarithm of a fraction is to use the bar notation. For example $\log 0·5 = \bar{1}·6990$, but in working out a sum like the one above it is best to avoid this notation and to write it out as $-1 + 0·6990 = -0·3010$. If you do not do this you will get into trouble when you come to multiply by 2·5.

This slight digression into mathematics is not essential. All one has to remember is that a magnitude is a number defining brightness: that the zero of the scale is arbitrarily selected; that larger numbers correspond to fainter stars; that first magnitude stars are the few brightest in the sky, and that stars of magnitude 6·0 are just visible to the naked eye. Further that a difference of 5 magnitudes in the brightness of two stars corresponds to a ratio of brightness of 100 to 1. (Put $m = 5$ in the formula and L becomes 10^{-2} or $1/100$). It is an illustration of the value of mathematical modes of expression that the whole of the last paragraph is implied in the one formula we have given.

Vega, which we have taken as approximately the zero point of the magnitude system is not the brightest star in the sky. Stars brighter than Vega must have magnitude numbers smaller than zero, i.e. they must be assigned negative numbers. The brightest star in the sky is Sirius, which has a magnitude of $-1·58$, i.e. it is 4·3 times as bright as Vega. Canopus, the brightest star in the southern sky has a magnitude of $-0·86$ which makes it 2·2 times brighter than Vega and half as bright as Sirius. You can verify these ratios from the formula.

As we shall see later the planets Venus and Jupiter, whose brightness is variable with position relative to the sun and earth, are often brighter than Vega and go up to magnitudes $-4·3$ and $-2·3$ respectively. The magnitude of the full moon is about $-12·5$ or 100,000 times as bright as Vega. The magnitude of the sun is $-26·7$ or 48,000 million times the brightness of Vega. Among miscellaneous facts which are of some interest are the following: a standard candle is as faint as a first magnitude star when it is at a distance of half a mile. An electric lamp of 20 candle power is reduced only to the sixth magnitude, i.e. is just visible, at a distance of about 20 miles, if absorption of light by the air is neglected. This is a forcible comment on the necessity of keeping strict black-out regulations during war time. A change of one-tenth of a magnitude is roughly the same as a ten per cent change in the brightness of a star or a point of light. If a source of light is moved to twice its original distance it becomes 1·5 magnitudes fainter.

Constellations

When one looks at the sky one sees a jumble of stars which, however, seem to sort themselves out into groups or constellations. The form of each constellation is merely a result of the accidental arrangement of stars in space, and has no fundamental significance. In ancient times, shepherds and others who were forced by their occupation to be out of doors at night, saw in these haphazard groups the figures of gods and animals, and a folklore has grown up around the constellations. They have received, or have inspired, at any rate in the northern sky, the names of characters of the classical legends. A large proportion of the southern constellations was first named two centuries ago by the French astronomer Lacaille, then on an expedition to the Cape of Good Hope. It must be admitted that only the eye of faith can see in many of the constellations the likeness of the object which they are supposed to resemble. These groups have been retained with relatively few modifications as the basis of modern star nomenclature.

The boundaries of the constellations have been fixed by international agreement. The areas of the various constellations differ widely, ranging from between 50 and 100 square degrees for small constellations such as Sagitta, Equuleus, Crux and Scutum to very large constellations, such as, for example, Hydra, which straggles over nearly seven hours of Right Ascension.

The bright stars in each constellation are named, almost always in order of decreasing brightness, by the letters of the Greek alphabet. Thus, the brightest star in the constellation the Lyre, is alpha Lyrae, but it also has the name Vega. Quite a number of the brighter stars have special names, very often of Arabic origin, in addition to their specification by a Greek letter and a constellation name. The system is somewhat similar to specifying a man, either by his name, or (allowing one man to one house) by the number of his house and the street in which it stands. The Greek alphabet suffices for most of the brighter stars, but when these letters are used up, numbers may be assigned. Thus there is a famous star, 61 Cygni, a relatively faint star in the constellation Cygnus, the Swan, which was the first star in the sky whose distance was determined; in the last few years it has also been suggested that this star possesses at least one planet.

Finally when these numbers become inconvenient, faint stars below the limit of visibility are often designated by the name of the star catalogue in which they appear, plus the number in the list. Typical names are Groombridge, Lalande (the names of compilers of star catalogues), H.D. (the name of a catalogue produced by Harvard College Observatory: the letters stand for Henry Draper), C.P.D. (standing for Cape Photographic Durchmusterung) and so forth.

Stars which are of variable brightness are denoted by capital letters beyond R and the constellation name. Examples are R Coronae Borealis, U Geminorum, etc. When the variables in a constellation are numerous enough to exhaust the alphabet, double capitals are used.

In the appendix will be found a list of the positions, constellation names, and Arabic names of some of the brighter stars in the sky; a list of constellations; and the Greek alphabet.

The star maps divide the sky into six regions—the northern cap (50 degrees to 90 degrees north declination); the southern cap (50 degrees to 90 degrees south declination) and four equatorial regions (declinations 50 degrees north to 50 degrees south) from 0h to 6h, 6h to 12h, 12h to 18h and 18h to 24h right ascension.

The stars on these maps (Maps 2, 3, 4, 5) are on the meridian at midnight in the periods from the end of September to the end of December; the end of December to the end of March: the end of March to the end of June: and the end of June to the end of September, respectively.

The Northern Cap (Map 1).

No star in this group is visible south of 40 degrees south latitude. North of this parallel they begin to be visible as stars which rise and set. On the equator all stars of this group are visible as stars which rise and set. North of the equator an increasing proportion, extending southwards from the north celestial pole, become circumpolar, until, by latitude 40 degrees north, all the stars of this group are circumpolar.

The most prominent constellations are Ursa Major, Ursa

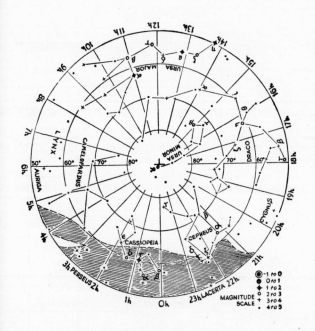

Map 1
THE NORTHERN CAP

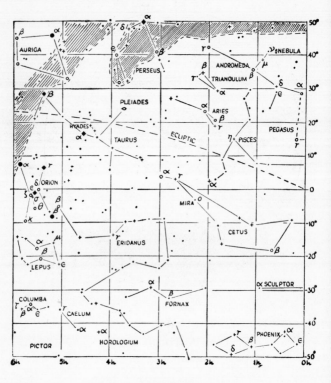

Map 2

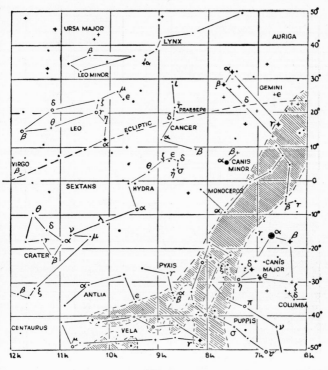

Map 3

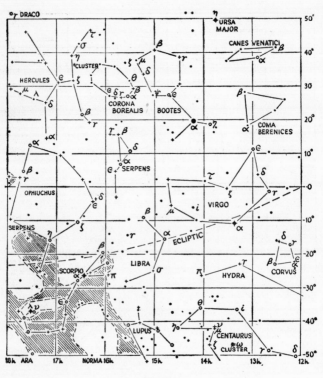

Map 4

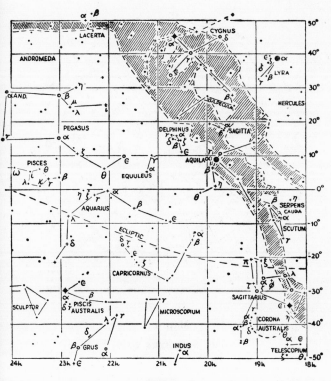

Map 5

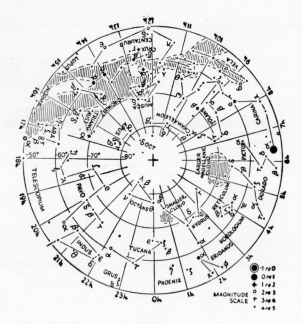

Map 6

THE SOUTHERN CAP

Minor and Cassiopeia. Less easily recognisable are Camelopardus, Cepheus and Draco.

Ursa Major. (The Great Bear, The Plough, the Big Dipper,
Charles' Wain, etc.) the most easily recognisable constellation
in the sky, lying between 8h and 14h R.A. and between declinations 50 degrees to 70 degrees north. There are seven principal
stars in the constellation, which is on the meridian at midnight in March, arranged in the familiar shape shown in the
map. The two bright stars α (Dubhe) and β are almost
exactly 5 degrees apart and the line joining them runs due
north and south, pointing to the pole star. The central star
of the " handle " Mizar (ζ) is double, having a companion
just distinguishable by the naked eye. In quite a small
telescope Mizar itself is also seen to be a double star.

Ursa Minor. Polaris, the Pole Star, α Ursa Minoris, is
1 degree away from the true pole and is the beginning of a
line of four faint stars of the fourth and fifth magnitudes
leading down to a rectangle of stars about 3 degrees by 5 degrees
the bottom two of which, β and γ, are the next brightest
stars in the constellation.

Cassiopeia, lying directly on the other side of the pole
from Ursa Major and Ursa Minor is at declination 60 degrees
and lies between R.A. 0 hours and 2 hours. The constellation
is in the form of a letter W about 12 degrees from tip to tip.
This is the chair of Cassiopeia. *Cepheus* lies above and to
one side of Cassiopeia. The most famous star of this constellation is Delta Cephei, a variable star. It varies with clockwork regularity, and is the prototype of a whole series of
variable stars, the Cepheid variables. Its variability has
been known since the 18th century (see Chapter V). *Draco*
and *Camelopardus* are long straggling constellations without
any very conspicuous form. Draco starts on the boundary
of Cepheus and runs in a semicircle round the pole to end as
a long tail between Ursa Major and Ursa Minor. Camelopardus lies on the opposite side of the pole between Ursa
Major and Cassiopeia.

The Autumn Stars (0h–6h) (Map 2).

The sun is at R.A. 0h in March and hence these stars are
on the meridian at midnight in the northern autumn, when
the sun is opposite to them in the sky. The path of the sun
(the ecliptic) is shown as a dotted line passing through the
Vernal Equinox (0h R.A., Declination zero).

This part of the sky contains a number of very prominent
constellations. In the northern half are Perseus, Auriga,

Triangulum, Aries, Taurus, Andromeda, Pisces, and the northern half of Orion. The southern half contains a number of less well-defined constellations, including Lepus, Columba, Caelum, Cetus, Fornax, most of the long straggling constellation Eridanus, and the remainder of Orion.

Perseus at Dec. 50 degrees R.A. 3h is a constellation consisting of a curved line of four bright stars about 10 degrees long, running up towards the W of Cassiopeia. Between the two constellations lies a fuzzy patch, the famous double cluster in Perseus which, even in a telescope of very moderate power is revealed as a star cluster containing many thousands of stars. β Persei is the variable star, Algol, mentioned in Chapter V. *Auriga* lies some twenty degrees to the east of Perseus. Auriga (the charioteer) is a constellation whose four brightest stars form a slightly irregular quadrilateral, with a fifth bright star (which actually belongs to Taurus, the neighbouring constellation) forming a pentagon. The brightest of these stars, Capella, is readily recognisable by its brilliance, its yellow colour, and by the fact that it is flanked by a triangle of faint stars. Capella is one of the best observed stars in the sky. In colour and temperature it is very similar to the sun, although intrinsically far brighter. The star is actually a very close double star, and the orbital motion, velocities and other characteristics of the pair are well determined.

Below Auriga and Perseus comes the constellation *Taurus*, the Bull. The principal star is Aldebaran, red in colour, lying at one vertex of a V of faint stars, a group known as the Hyades. Between Taurus and Perseus lie the group known as the Pleiades, about seven stars of the 3rd–5th magnitudes being visible to the naked eye. Other fainter stars can be glimpsed under good conditions. The stars are, as they appear to be, a group situated relatively close to each other in space, much as are the constituent stars of the Perseus cluster, but there is neither the same density nor range of brightness among the stars of the Pleiades. In some ways the Pleiades is a cluster very similar to the cluster including the constellation Ursa Major which also forms a group of stars relatively close to each other in space. However the Ursa Major cluster is nearer to us, so that the association of the constituent stars is by no means as obvious as it is in the case of the Pleiades. The Pleiades stars are, as has been proved by long-exposure photographs, surrounded by clouds of tenuous gas.

Andromeda in the same declination as Perseus, consists

of a line of four bright stars leading up from the corner of the " Great Square " of *Pegasus* (see below).   The corner star of the square is the second magnitude star Alpheratz (α Andromedae), while the star at the further end of the line is γ Andromedae a beautiful double star with contrasting colours resolvable in a quite small telescope.   To the west of ν Andromedae is a faint patch of luminosity.   This is the great nebula in Andromeda, the brightest and nearest of the so-called extragalactic nebulae, and a structure similar in size to our own Milky Way.

Triangulum lies beneath Andromeda, and further south, forming a smaller obtuse-angled triangle is *Aries*.   South of this lies *Cetus*, the whale.   Easily the most striking constellation in this quadrant is Orion, 20 degrees from north to south and 10 from east to west, lying right on the celestial equator. It is outlined by an irregular quadrilateral of stars.   In the north, the brightest star in the constellation (α Orionis) is Betelgeuse, bright red in colour.   This star is relatively cool, and radiates most of its radiation not as visible light, but as light of slightly longer invisible radiation (infra-red radiation). If we could see this radiation (and various instruments are available which can detect it), we should find this to be the brightest star in the whole sky.   By various methods it is possible to estimate the true diameter in miles of certain stars, one of which is Betelgeuse.   Betelgeuse is so large that, were it located where the sun is, the orbit of the earth would be inside.   The companion star at the northern end of the quadrilateral is Bellatrix, a bright blue star.   At the centre, lying on the equator, is a line of three bright stars, the belt of Orion, with, immediately below, a line of three fainter and fuzzier looking stars, forming the sword hanging down from the belt.   Southern hemisphere readers will of course see this constellation upside down, with the sword pointing upwards.   The line of the belt, extended south-east, leads to Sirius, the brightest star in the sky (see below) while in the opposite direction the line leads to Aldebaran, in Taurus, already mentioned.

The more westerly of the two lower stars of Orion's quadrilateral is blue and has the name Rigel.   The whole of the constellation of Orion has been shown to be surrounded by a cloud of glowing gas.   The densest portions include the cloud round the central " star " of the sword (θ Orionis), a structure usually called the Great Nebula in Orion.   This " star " is, in quite small telescopes, seen to be quadruple. The four stars are known as the Trapezium.   Throughout the Orion constellation there is a wealth of structure produced by clouds of gas and of dark dust intermingled with the stars. The scale of these structures is enormous: the distance of the earth from the sun is quite insignificant in comparison. Photographs of these structures, such as the Horse's Head, are most impressive, but, like most astronomical photographs reproduced in books, they show a far higher contrast than appears to the eye looking through a telescope. First, long-exposure photography with very fast plates has been used to increase the brightness of the faint gas clouds, and often, special light filters and plates which pick out the particular colours in which the emitted light is most intense have been used. Finally, it is almost always necessary in preparing prints for publication artificially to increase the contrast still further by various devices of intensification.

The remaining constellations are less striking than those already described, and it is hardly advisable to try to identify them when first finding one's way about the sky. The two most prominent are Phoenix and Columba.

The Winter Constellations (6h–12h). (Map 3).

The most striking of these are Gemini (the twins), Leo (the lion), Canis Minor, and Canis Major (the lesser and the greater dog). Others in this region are Leo Minor, Cancer (the crab), Sextans (the sextant) part of Hydra, Monoceros (the unicorn), part of Puppis, Pyxis, Antlia, and Crater (the bowl). This is the part of the sky lying roughly south of the ploughshare part of Ursa Major. The three prominent constellations, Gemini, Canis Minor and Canis Major lie on a line roughly north and south. Gemini, in north declination 30 degrees is next to Auriga. Its two brightest stars are Castor and Pollux just five degrees apart, Castor the more northern is α Geminorum and fainter than Pollux, β Geminorum. This is a case where alpha of a constellation is not the brightest star in it. Castor, is a yellow star and in a telescope is seen as double. *Canis Minor* just over 20 degrees south of Gemini also has a prominent pair of stars very similar in separation and on an almost parallel line to Castor and Pollux. Confusion between the two constellations is unlikely. In Canis Minor the more northerly star is a good deal fainter than the more southerly (Procyon) as against the more nearly equal brightness of Castor and Pollux. If we continue south along the line from Castor to Pollux for about 10 degrees we come to the two fairly faint stars δ and γ Cancri. Slightly to the west of this pair we find a fuzzy patch. This is another cluster of stars, Praesepe (the beehive) a fine sight in a small telescope. Large numbers of faint stars from the 7th to the 8th magnitude are to be seen.

The brightest star in *Canis Major* is Sirius, to be located from the belt of Orion as already described. Sirius is the brightest star in the sky, and of a blue colour; the twinkling produced by irregular air currents in the earth's atmosphere, may cause it to seem to flash a variety of colours. All stars twinkle more or less, according to the atmospheric conditions which prevail. The effects of variations of brightness and colour and small shifts of position are produced entirely by the earth's atmosphere and have nothing to do with the stars themselves. The effects, present for all stars, are merely more easily seen in the case of a very bright star such as Sirius.

Sirius, the " dog star " is bright enough to be seen even when close to the sun. The sun passes Sirius in its passage round the ecliptic early in July, in the heat of summer in the northern hemisphere. These are the dog days, of reputed hot weather, and are so called because the constellation Canis Major is then seen near the sun. The name has nothing to do with the idea that the weather becomes so hot that dogs go mad. The sun and Sirius rise simultaneously early in July, the expected date of arrival in lower Egypt of the Nile flood on which the whole life of the country depends. The necessity of accurately predicting this date was one reason for the early development of astronomy in Ancient Egypt.

Although Sirius is the brightest star in the sky and has been well observed since astronomy started, it was not until 1862 that it was discovered that it consists of a pair of stars, one much fainter than the other, which move around one another in space under their mutual gravitational attraction. The companion star Sirius B can only be seen in very large telescopes, not so much because Sirius B is particularly faint it is of magnitude 8) but because the strong light of Sirius A blots it out. Sirius B, insignificant though it is, is one of the most interesting stars known to astronomy. Modern investigation has shown that, though it has only about 4 times the diameter of the earth, it has a mass comparable with that of an ordinary star—that is something like 300,000 times the mass of the earth. The density of its material is thus about 30,000 times that of water—one ton in a matchbox as the phrase goes—and the problem of how matter can exist in such a close packed state has thrown interesting side-lights on modern ideas about atomic structure. Detailed

investigation of the light from Sirius B which is affected by
the intense gravitation produced by so concentrated a mass,
afforded one of the three crucial proofs of Einstein's theory of
relativity.

The last of the prominent constellations in this group is
Leo, lying on the ecliptic. The brightest star Regulus lies
at the foot of a curved sickle-shaped line of stars—this group
is often called the Sickle in R.A.F. parlance—but the full
constellation includes a triangle of stars to the east forming
the hind quarters of the lion. The brightest of these is
β Leonis or Denebola.

The Spring Stars (12h–18h). (Map 4).

In the quadrant from 12 to 18 hours Right Ascension the
principal constellations are Virgo, Boötes, Hercules and
Scorpio (the scorpion). Others, less easily recognised, are
Canes Venatici (the hunting dogs), Coma Berenices (Berenice's
hair), Corvus (the crow), part of Hydra and Centaurus,
Corona Borealis (the northern crown), Serpens, Ophiuchus
and Libra (the scales). For an observer in the northern
hemisphere the best way of picking up the principal land-
marks is to extend downwards the curved line of the handle
of the Plough. Going south along this line one comes
first to Arcturus (α Boötis), the red star in declination 20
degrees which is the brightest in this constellation. Going
on down another 30 degrees brings us to the bright blue
star, Spica, the brightest star of the constellation Virgo.
Canes Venatici lies beneath the handle of the Plough its
brightest star being the third magnitude star, Cor Caroli
(α Canum Venaticorum). The curious name " Cor Caroli ",
Charles' Heart, is said to have originated at the time of the
restoration of Charles II when this star is alleged to have
brightened with joy. *Coma Berenices* lies below this and
above Virgo, and consists of a mass of fourth, fifth and sixth
magnitude stars suggesting the wavy tresses which give the
constellation its name. *Corvus* is a small quadrilateral with
sides of about 5 degrees lying south of Virgo. Between Boötes
and Hercules is the appropriately named *Corona Borealis*,
a tiara or hoop of third and fourth magnitude stars. *Hercules*
lies between Corona and the brilliant blue-white star Vega
in the next quadrant of the sky. Its brightest star is only of
the third magnitude and the constellation is best recognised
by a group of five other third magnitude and one other fourth
magnitude star arranged in two lines running roughly north
and south. The six stars are in pairs, the centre pair being

a little closer than those north and south of it. On the line joining the top right-hand pair (η and ζ Herculis) can be glimpsed with the naked eye the fuzzy patch which is the famous globular cluster in Hercules, a highly symmetrical cluster of stars centrally condensed, and forming a group typical of many other fainter objects scattered around the sky.

South of the equator is *Scorpio* whose brightest star, Antares, is a brilliant red first magnitude star to be seen low on the southern horizon on summer nights in England. The curving line of stars which gives the constellation its name will hardly ever be seen from northern Europe. They may just be glimpsed but only on a night when the sky is perfectly clear right down to the horizon and when there are no obstructions. The line imitates well the tail of a scorpion curving up and over the back to sting an attacker. Antares is a double star and has a fainter blue companion three seconds of arc away. It is a fine sight in a telescope and the contrasting colours are very striking. Antares itself is somewhat similar to Betelgeuse, being in fact a cool star of a very large size.

The Summer Stars (18h–24 h). (Map 5).

This quadrant contains some of the most striking constellations in the sky. The most prominent are Cygnus (the swan) Lyra, (the lyre), Aquila, (the eagle), Sagittarius (the archer), Pegasus (the winged horse), and Piscis Australis (the southern fish). Less easily recognisable constellations are Vulpecula, Sagitta, (the arrow), Delphinus (the dolphin), Equuleus, Aquarius, Capricornus, Microscopium, Scutum and Corona Australis.

Cygnus, sometimes called the northern cross, is a large well-marked constellation with a line of four stars 25 degrees in length, of the 1st, 2nd, 4th and 3rd magnitudes marking the upright of the cross. Two third magnitude stars 15 degrees apart mark the cross bar. To the west is Lyra whose brightest star Vega is just circumpolar in Britain. The remaining prominent stars of this small constellation form two pairs, the more northerly a pair of fourth magnitude stars 2 degrees apart, and parallel to these and 5 degrees south of them is the second pair, separated by the same distance, this time of third magnitude stars. This last pair are β and γ Lyrae. ε Lyrae is a famous star, the " double double ", a star just double to the naked eye, each component of which is telescopically double

Further south is *Aquila*, the brightest star, Altair, being

c

flanked at a distance of 2 degrees on either side by 3rd and 4th magnitude stars forming the wings of the eagle. Close by is the appropriately named *Delphinus*, a little kite with a tail drawn in 3rd and 4th magnitude stars, which, curved as it is, suggests the arched back of a dolphin leaping out of a wave. *Sagitta*, the arrow, just north of Aquila, is a straight line of rather faint stars dividing on the west to form the feathers of the arrow. The rather numerous bright stars of *Sagittarius*, south of Aquila, form the rather complex shape illustrated in the map.

The Great Square of Pegasus, already referred to, measures 15 degrees by 15 and the direction of its western vertical side takes one down to the bright star Fomalhaut (α Piscis Australis) in declination 30 degrees south. Scutum (the shield) and Corona Australis (the southern crown) both bear a considerable likeness to the objects which they are supposed to represent.

The Southern Cap (Map 6).

The principal constellations are Carina, Crux (the southern cross) and Centaurus (the centaur). There are also part of Eridanus, Hydrus, Horologium, Reticulum, Mensa, Dorado, Pictor, Volans, Puppis, Chamaeleon, Vela, Musca, Circinus, Triangulum Australis, Apus, Lupus, Norma, Ara, Octans, Pavo, Telescopium, Indus, Tucana and Grus.

Crux, the southern cross, is south of Virgo and should be recognised without fear of confusion by reason of its small size and the brightness of its four stars. A fifth star of the fourth magnitude lies on one edge of the kite shape formed by the other four stars. Immediately next to the Cross lies the conspicuous black patch, the Coal Sack, in reality a cloud of opaque dust and gas hiding more remote stars from view. At 2h greater Right Ascension are the two brightest stars of *Centaurus* (α and β Centauri) five degrees apart, pointing towards the Cross. The direction of the axis of the Cross (γ Crucis to α Crucis) and the perpendicular bisector of the join of α and β Centauri intersect at the south pole of the heavens, but there is no bright star to mark it. The nearest star in space to the sun is Proxima Centauri a faint star in this constellation. Alpha is almost as near.

Canopus, the second brightest star in the sky is in Carina, separated by some distance from the main part of the constellation. It is almost due south from Sirius. The remaining very bright star in all this region is Achernar, brightest star—almost the only bright star—in the straggling constellation of Eridanus. It is at south declination 60 degrees directly opposite Crux and Centaurus over the pole of the heavens.

Two of the most striking features of the southern sky are not constellations at all but very large and complex clusters of stars. These are the Larger and Smaller Magellanic Clouds in Dorado and Tucana respectively. They appear to be, what they probably are, additions to or satellites of the Milky Way. To the naked eye the Large Cloud seems about six degrees across, the Smaller Cloud about half as large. Both the Magellanic Clouds can be seen in the telescope and in photographs, to include vast numbers of stars, clouds of gas, and, more especially in the Large Cloud, clouds of obscuring dust. The nebula known as 30 Doradus, enmeshed in the Large Cloud, is a good object for a fairly small telescope. Close to the Small Cloud in the sky, but at only a fraction of its distance from us in space, is the globular cluster, 47 Tucanae, second only to the globular cluster, ω Centauri. In the northern sky the brightest globular cluster is that in Hercules. (see p. 54).

A word may be added about the constellations of Argo and Serpens, both of which have suffered " political " change at the hands of the international co-ordinating body of astronomy, the International Astronomical Union. This body had to make a rational system out of the stellar profusion provided by nature and the mental confusion provided by legend. All constellations now have borders running along parts of parallels of declination and meridians of right ascension like the states of the U.S.A. Argo, the celestial ship, a sprawling mass of bright stars, was long since divided into Puppis, the poop, Vela, the sails, and Carina, the keel.

This has left the ordinary nomenclature somewhat confused since the Greek letters employed referred to the old constellation Argo and not to the separate parts into which it is now divided. Serpens has been split into two entirely separated parts like the County of Flint, Serpens Caput (the Serpent's head) at 15h–16h, and Serpens Cauda (the Serpent's tail) at 18h–19h.

There is, in addition to the stars, a further, very striking permanent feature of the sky. This is the Milky Way, an irregular band of faint glowing luminosity which encircles the whole sky (shown hatched on the maps). Starting at Cassiopeia we can trace it through Auriga, along between Orion and Gemini, through a point just north of Sirius, down to the Southern Cross and Centaurus; then along through the curled tail of Scorpio northwards through Aquila and Cygnus and back to Cassiopeia. From Centaurus north to Cygnus it is divided by a dark lane, now known to be due to dust and other obscuring matter in space lying in front of the Milky Way. To anticipate what will follow: the Milky Way, sometimes known also as the Galaxy or Galactic System, is the star system of which our sun is a member. Its general diffuse radiance can, with telescopic aid, be shown to be due to myriads of stars each individually too faint to be distinguished, but providing in the mass a general glow of light. Although in photographs taken with large telescopes the stars in the Milky Way seem to be crowding together so that they almost seem to touch, this is an illusion. The stars of the Milky Way are in fact, even in its densest parts, separated from each other by distances almost as large as those which lie between the stars nearer to the sun, and if we were to be transported to any part of the Milky Way the number of bright stars in the sky would probably not be increased many times.

Research has shown that, were we able to get right outside the system and to view it as a whole we should see that the stars of the Milky Way defined a shape much like the currants in a currant bun of high fruit content. The system is round when looked down upon, but flattened when seen in profile. The currants are more concentrated towards the centre of the bun, and the sun, an insignificant fruit, lies fairly far out from the centre in the plane where the bun might be sliced for buttering. All our view of outer space is had from this position, from which we look out through a cloud of stars in the foreground. Looking in any direction in the plane of the butter layer we see many stars—the Milky Way as we know it. Looking at right angles to this direction we find in our way only the thin layer of the half bun so that very few stars are seen. It is this concentration of the stars towards the centre of the bun, and towards the plane of the butter layer which produces the effects which we see. This is the picture on the grand scale: in detail there are many irregularities—extra concentrations of stars in the form of clusters; clouds of glowing gas such as the nebula in Orion; clouds of obscuring dust like the division in the Milky Way already mentioned, or like the " Coal Sack ", the dark area near the southern cross. (This is the unshaded area on Map 6.)

Problems of the structure of the Milky Way occupy a leading position in modern astronomical research. But this is a grand scale problem. The kind of astronomy which one can teach oneself must inevitably start with things much nearer home. We must reduce our scale of ideas something like 10,000 million times from the scale of the Milky Way, and turn next to the planets which move round the sun.

CHAPTER III

THE PLANETS

Planetary Motion

Everything there is to be known about the motions of the planets round the sun and their relative distances from it can be deduced from observations of the positions of the planets against the star background. Over a period of several centuries, this is the method by which modern knowledge of the solar system has been gained and anyone with sufficient patience and skill could repeat this procedure. Life is too short for each of us to do this, so we shall adopt the reverse procedure, that is, to set out the modern knowledge and to show what phenomena this implies. You may then fairly easily verify for yourself that these phenomena actually occur.

As we said at the end of the last chapter, the sun and its attendant planets, the solar system, are quite insignificant in size by comparison with, for example, the Milky Way. The brilliance of the planets in the sky and their striking appearance are due solely to their very close promixity in space to us and to the sun. Even though, as judged by the size of the earth, the diameters of some of the planets are very great, and their distances from the sun very large indeed, the planets are of a very local and parochial interest. The solar system itself, though possibly not unique, is certainly the only planetary system which we shall ever be able to study at all closely. For us, a planet will mean one of the sun's planets, a relatively small piece of solid matter moving round the sun in a path or *orbit*, owing its light and surface heat to the sun, and held in that orbit by the attraction which the sun exerts on the planet.

Let us begin with a model: a small weight on the end of a string. Whirl this around your head. You can feel that there is a tension in the string, and you know that you must keep pulling on the string as you whirl the weight round, for, if you did not, and let the string go, the weight would fly off at a tangent. When the weight is going round at a steady pace, the pull which you exert on it is constant. When the weight moves at a faster steady pace the pull in the string increases. It is this pull which constrains the weight to move in a curved path, a path in which the direction of the motion of the weight is continually turning. This force is often called incorrectly the centrifugal force. If the weight in the model represents a planet, then the pull exerted by the string represents the gravitational attraction of the sun acting on the planet. There is no longer any visible connection, but the force is there nevertheless, and is of precisely the same kind as the gravitational force which holds us on the earth. In the empty space between the planets there is no other force acting and even this gets rapidly weaker as one goes away from the sun, although it never, strictly speaking, vanishes completely. To make a body travel in a curved path there must be a force acting across the direction of motion of that body and in the case of the planets the force is provided by the gravitational attraction of the sun. Where gravity is strongest, i.e. close to the sun, these paths will be most curved, and the planets moving in them will be moving faster, so that just as

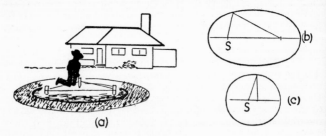

Figure 21

in the case of the weight on the string there is an increased centrifugal force to balance the larger force of gravity. At more remote positions, a planet will move more slowly.

Thus, a planet near the sun will be moving fast in its orbit: Mercury does 29·7 miles per second: the earth, at a greater distance does 18·5 miles per second. Jupiter, five times as far from the sun as the earth travels at about 8 miles per second. The orbits of all planets are examples of the type of oval curve known as an ellipse. (Figure 21.) Every ellipse has two special points on the longer axis of symmetry which are called the foci of the ellipse. The simplest everyday example of an ellipse and its foci is provided by the practice of gardeners in laying out oval flower beds. They stick two pegs into the ground and put a loop of string over them and over a third movable peg. This third peg is pushed into the ground at the various positions to which it can reach when the string is fully stretched. These points all lie on an ellipse and the two fixed pegs are the foci. If the pegs are close together the ellipse is almost a circle, and if the two pegs are coincident, clearly the movable peg traces out a circle round the common centre. If, compared with the size of the loop of string the two fixed pegs are placed far apart, the ellipse is very long and thin—or, in astronomical parlance, is very eccentric, or has a high eccentricity.

A consequence of the laws of gravity and of mechanics is that all planetary orbits are ellipses, but most of them are in fact of small eccentricity, that is, almost circular. The sun is at one focus of all these elliptical orbits, and, for orbits of small eccentricity, such as those of the planets, the sun lies almost at the centre of the almost circular orbits. As we have seen, when a planet is near that part of its orbit where it is closest to the sun (near *perihelion*) it is moving faster than when it is most remote from the sun (near *aphelion*). We have already met this phenomenon of variable orbital velocity of a planet in connection with the orbital velocity of the earth and the equation of time.

The Solar System

The average distance of the earth from the sun is one of the standard measures of length in astronomy. It is called the *astronomical unit*, and according to the latest determination measures 93,005,000 miles. We shall indicate later how the determination of this length is carried out. The following table gives the names and certain data for the planets.

The Planets

Name	Distance from Sun (millions of miles)	Distance from Sun (in astron. units)	Orbital Period (years)	Period of Rotation	Diameter (miles)	Velocity in orbit (miles per second)
Mercury	36·0	0·39	0·241 (= 88d)	? 88d	3,100	29·7
Venus	67·3	0·72	0·615 (= 225d)	? 30d	7,680	21·7
Earth	93·0	1·00	1·000	23h 56m	7,927} 7,900}	18·5
Mars	141·7	1·52	1·88	24h 37m	4,220	15·0
Jupiter	483·9	5·20	11·86	9h 50m	88,700} 82,800}	8·1
Saturn	887·2	9·54	29·46	10h 14m	75,100} 67,200}	6·0
Uranus	1784·0	19·19	84·02	10h 45m	31,800	4·2
Neptune	2797·0	30·07	164·8	15h 48m	27,800	3·4
Pluto	3670·0	39·60	247·7	?	3,600	2·9

The Satellites

Planet	Satellite	Distance from Planet (thousands of miles)	Diameter (miles)	Period d.	h.	m.
Earth	Moon	239	2,160	27	7	43
Mars	Phobos	5·8	10		7	39
	Deimos	14·6	5	1	6	18
Jupiter	V	113	100		11	57
	I Io	262	2,060	1	18	28
	II Europa	417	1,790	3	13	14
	III Ganymede	666	3,070	7	3	43
	IV Callisto	1,171	2,910	16	16	32
	VI	7,146	75	250	16	
	X	7,301	12	260		
	VII	7,301	25	259	19	
	XII	13,049	12	625		
	XI	13,981	15	696		
	VIII	14,602	25	738	22	
	IX	14,726	14	755		
Saturn	Ring System	44–85	—	—		
	I Mimas	115	320	0	22	37
	II Enceladus	148	370	1	8	53
	III Tethys	183	750	1	21	18
	IV Dione	235	810	2	17	41
	V Rhea	328	1,120	4	12	25
	VI Titan	760	3,110	15	22	41
	VII Hyperion	922	250	21	6	38
	VIII Iapetus	2,214	750	79	7	56
	IX Phoebe	8,047	190	550	11	
Uranus	Miranda	80	?	1	9	56
	Ariel	119	370	2	12	39
	Umbriel	166	250	4	3	28
	Titania	273	620	8	16	56
	Oberon	364	500	13	11	7
Neptune	Triton	220	2,490	5	21	3
	Neried	3,728?	190	500		

Some planets are flattened at the poles due to rapid rotation. Where two diameters are given the smaller is the polar, the larger the equatorial. Note the immense size of interplanetary distances compared with the dimensions of any planet.

Charting the Solar System

Now let us consider the appearance of the solar system as seen from the earth. A study of the table shows that two of the planets, Mercury and Venus, are always closer to the sun than the earth is. They have shorter orbital periods and move faster in their orbits. Suppose we consider the motions of the earth and of Venus starting at a moment when the sun, Venus and the earth are all in line. Clearly Venus will gain on the earth so that, whereas at the beginning it was in line with the sun, at a later time it will lie out to the west of the sun. (Figure 22.) From the earth, Venus will then day by day be seen to move further west of the sun, so that before dawn, before the sun has risen, and before Venus has been lost to view in the brightness of the daytime sky, the planet will appear as a "morning star". The angular distance between Venus and the sun will not however

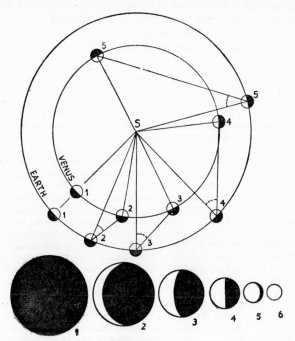

Figure 22

Positions of Earth and Venus in their orbits, and corresponding telescopic appearance of Venus.

go on increasing indefinitely. A time will come when Venus is so far ahead of the earth that it will, as it were, be leading the earth in the race, right round the bend of the course. The elongation of Venus from the sun will increase to a maximum and then start to decrease, and as the figure shows, this will happen when the line from the earth to Venus just touches

c*

the orbit of the latter. If we look at the geometry of the figure we see that, at this moment, the line from the earth to Venus is at right angles to the line from Venus to the sun. We can measure directly the angle between the direction to Venus and the sun at this moment, and we shall find a value of about 46 degrees. Elementary trigonometry then enables us to deduce the ratio of the distances from the earth to the sun and from Venus to the sun. The relation is

$$\frac{\text{Distance of Venus from sun}}{\text{Distance of Earth from sun}} = \text{Sin } 46° = 0.72.$$

In other words, by the simplest of observations we can compare the ratio of the radii of the orbits of Venus and the earth round the sun.

After this moment of greatest elongation Venus appears to move in towards the sun, finally passing behind it. A little later it reappears on the other side of the sun as an " evening star ", reaches a maximum elongation on that side of the sun—the value of this maximum angle again being about 46 degrees—and then moves back towards the sun until, having gained a whole revolution on the earth, Venus once more crosses the line joining the earth to the sun.

The general characteristics of the motion of Mercury are much the same except that the maximum value of the angular elongation is half the corresponding value for Venus or about 23 degrees. These angles give the maximum distances ahead or behind the sun to which these two planets can attain. In terms of the axial rotation of the earth they represent 3 hours and 1½ hours respectively. That is to say that Mercury never sets more than about 90 minutes after the sun or rises more than 90 minutes before it : Venus never sets more than about 3 hours after the sun or rises more than 3 hours before it.

We have been describing the characteristics of the motion of a planet whose orbit is interior to that of the earth. These are, that the planet in question always lies fairly near to the sun in the sky, being sometimes on one side of the sun and sometimes on the other. From the maximum value of the separation of the plane from the sun it is possible to deduce the distance of the planet from the sun in terms of the astronomical unit.

Phases of Inferior Planets
Both Mercury and Venus like all the other planets owe all their light and heat to the sun, and shine solely by the light reflected from their surfaces. Each planet is, therefore,

at any moment half in darkness and half illuminated by the sun. The brightness of a planet will depend, among other things, on how much of this illuminated half we can see. Suppose, for example, Venus is on the opposite side of the sun from the earth. Then the planet will appear very close to the sun, and, if we could see it, we should see its illuminated half presented to us. On the other hand, this illuminated disc will appear very small because the planet is then at its maximum distance from us. When Venus is closest to us, that is, is almost in a line from the earth to the sun, but on the same side as the earth, the planet has its maximum apparent diameter but the side which is turned towards us is the shadowed one. In this case then, the planet will again appear relatively faint. When the planet is at its maximum elongation from the sun we shall see half of the illuminated side and half of the dark side, that is, in a telescope the planet will look like a tiny moon at first or last quarter. In this sort of position the planet will look very bright indeed, for although only half the disc is illuminated, the distance of the planet from us is not very great. It will be clear from what has been said that the inferior planets like Venus and Mercury show phases as the moon does, and it was one of the crucial tests of the theory of the solar system when first propounded, that these planets should show phases. The maximum brightness of these planets occurs when rather less than half the disc of the planet appears illuminated, that is when they show as crescents, for then the distance is still smaller than when they are at maximum elongation and the effects of distance and of proportionate illumination combine to produce maximum brightnesss.

The phases of Venus can quite readily be distinguished with the aid of a moderately good pair of field glasses, and, since the planet moves round the sun in a relatively short time the variation from night to night can easily be followed. (Figure 22.)

The Superior Planets

The planets exterior to the earth:—Mars, Jupiter, Saturn Uranus, Neptune and Pluto, of which the first three are naked eye objects, and the fourth an easy object for a small telescope or field glasses, all present only their illuminated surfaces to the earth. All these planets are far away compared with the earth, and are all roughly opposite to the sun when most conveniently visible at night so that no part of their dark surface is visible. An exception is Mars, which is 1·52 times as far from the sun as the earth is, which means that in suitable circumstances a small part of the dark side can be seen. When this happens Mars is seen gibbous as the moon is just before or just after full moon.

If we compare the motion of the earth with that of an exterior planet as we did in the case of the earth and Venus we find a different situation. Now the earth catches up on the slower moving exterior planet. Starting again from a position when the sun, the earth, and, say, Mars are all in line, we see that, at first, the line joining the earth and Mars swings in the opposite direction to the common motion round the

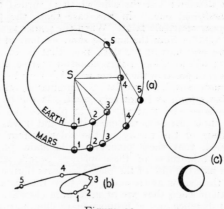

Figure 23

(a) Positions of Earth and Mars in space; (b) corresponding path of Mars on the sky; (c) appearances of Mars at positions (1) and (5).

sun. (Figure 23.) When however the earth gets so far ahead as to be going " round the bend " the line swings in the opposite direction. That is, as seen in the sky, the motion of Mars is first in one direction and then in the other. This same direction persists as the earth goes on gaining until it passes round the other side of the sun. When this happens it means that Mars has disappeared in the sun's brightness. Then, as the earth goes on, Mars reappears on the other side of the sun, still apparently going in the same direction against the background of the stars. Presently the earth begins to gain a whole lap on Mars, and, as it comes round the bend still catching up, the line from the earth to Mars reverses its direction of swing once more, i.e. Mars appears to change its direction of motion against the star background, and continues this reverse or " retrograde " motion until the sun, the earth, and Mars are once more in line. The effect of this is that Mars, and every other superior planet, appears each year to make a loop in its track against the sky background. It is possible to deduce from the size of this loop, just as in the case of Venus, how many astronomical units Mars is distant from the sun. What it comes to, then, is that by quite simple observations of angles it is possible to determine the distance from the sun of all the planets in terms of the astronomical unit; that is to say, it is easily possible to construct a scale model of the solar system. What is lacking is the length of the astronomical unit in miles: we know all the relative distances, but not the value in terms of the familiar terrestrial units of length.

Before considering this point, there is one note which must be added. We have spoken frequently of the earth, the sun and a planet being in line. This is usually only very roughly true. If the orbits of all the planets were in exactly the same plane, that is, if a true model could be made in which all the model planets moved on the same table top, then, each time one planet passed another the shadow of one would cross the other and there would be an eclipse. In fact, although all the orbits of the planets lie nearly in the same plane, the coincidence is not exact: the plane of the orbit of one planet is slightly tilted with respect to the orbit of any other planet, and eclipses are a rarity which occur only when the bodies concerned happen to be just at the right points in their orbits at the right time. Thus when Venus passes between the earth and the sun it usually does so just above or just below the sun: only very occasionally does Venus pass across the face of the sun, producing what is known as a transit of Venus.

The characteristics of the motion of external planets are thus that these planets may be seen anywhere round the ecliptic and are best seen when opposite to the sun, i.e. when they cross the meridian around midnight, and that each year, as a reflex of the earth's motion, they show a loop in their paths against the star background. The loop would be a back and fore zigzag if all the planetary orbits were in the same plane. The slight differences of plane already mentioned turn each zigzag into a loop. On the other hand the differences of orbital inclination are fairly small, so that all the planets are to be found in a fairly narrow band encircling the sky. This narrow band includes the ecliptic, the apparent orbit of the sun round the earth. This obviously must be so because

the track of the sun against the stars as seen from the earth
marks the plane of the earth's orbit round the sun, which, as
we have said, is close to that of all the other planets.

We have seen how to construct a scale model of the solar
system by determining the distances of the planets from the
sun in terms of the astronomical unit.   In the table we have
also listed the periods of the planets—the true time which each
takes to make one circuit round the sun.   The apparent
times are affected by the fact that the earth is moving.   How

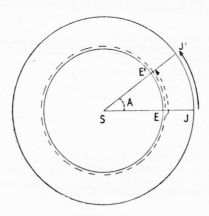

Figure 24

can we deduce what the orbital period of a planet is from the
interval between the times when the earth passes it as both
move round the sun ?

Synodic Period

Suppose the earth goes round the sun in a certain time,
say T_E days, while Jupiter goes round in a time T_J days.
Then if both start off from the same starting line, they will
be level again when the earth has caught up one lap on Jupiter.
Suppose that when the two planets are level again the common
line has advanced by A degrees (see Figure 24) and suppose
a certain number of days T has elapsed.   Then in this time T
days, Jupiter will have gone round through A degrees, but the
earth, having caught up a whole lap will have gone round
A + 360 degrees.   Now to make these two statements correct,

certain relations must apply. For Jupiter the period of T days must represent the proper proportion, $A/360$ of its orbital period, i.e.

$$\frac{T}{T_J} = \frac{A}{360}$$

For the earth

$$\frac{T}{T_E} = \frac{A + 360}{360} = \frac{A}{360} + 1$$

These equations simply express the facts that both for the earth and Jupiter the angle through which they have gone is proportional to the time elapsed, putting in the proper rates for each planet. We are not interested in A, so to get rid of it we take its value given by the first equation, and put it in the second. Now we get what we do want, which is the relation between T and the orbital periods of the two planets. The result is

$$\frac{T}{T_E} = \frac{T}{T_J} + 1$$

or, dividing all through T, and rearranging

$$\frac{1}{T_E} - \frac{1}{T_J} = \frac{1}{T}$$

T is what is called the *synodic period* of the two planets, the interval between successive meetings on the same line. For the earth and Jupiter taking $T_E = 1$ year and $T_J = 12$ years, we have $1/T = 11/12$ or $T = 12/11$ years. That is, Jupiter is opposite the sun as seen from the earth about every 13 months. In practice the calculation works the other way. Observing that Jupiter is opposite to the sun every 13 months we can work out what the true orbital period of Jupiter is. Thus, for example Mars passes the earth once every 780 days approximately. Putting T_M, for the period of Mars, in the relation given above we have

$$\frac{1}{365} - \frac{1}{T_M} = \frac{1}{780}$$

which gives $T_M = 687$ days. Notice that in the equation $\frac{1}{T_E} - \frac{1}{T_J} = \frac{1}{T}$ we can replace T_E and T_J by the periods of any two planets but the second one (with the minus sign) must refer to the one farther from the sun.

We have thus sketched the bare bones of the methods by which the distances of the planets from the sun in astronomical units, and their orbital periods can be determined. Knowing

the radii of the orbits we can fairly easily determine the
distance of each planet from the earth at any time. Then,
by noting the apparent diameter of the planet we can estimate
its size, still in astronomical units; if the planet has a moon,

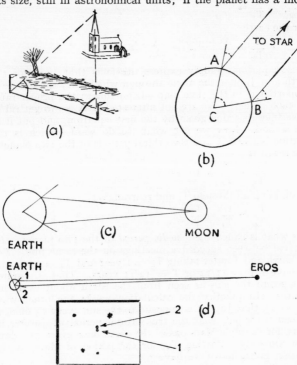

Figure 25

*Scale determination summarised in sketches: (a) is the surveyor's
method of determining the distance of an inaccessible object. In
(b) astronomical observations determine the angular distance AB.
Surveys from A to B on the earth's surface give AB in miles, and
hence the dimensions of the earth. In (c) this data, and the method
of (a) are used to find the distance of the moon in miles. In (d) the
same method is applied to find the distance of Eros, but this is so
great that when observations from positions (1) and (2) are compared,
there is only a small, but still measurable, shift in the position of
Eros as seen against the star background.*

we can estimate the distance of the satellite from the parent planet in astronomical units.

Fixing the Scale

To fill in the gap in this chain of deduction we indicate how the value of the astronomical unit in miles can be determined. Clearly, since we can make a scale model of the solar system, all that we need to do is to determine the distance between any two planets, at any time, in miles, and to compare this with the value computed in astronomical units. The difficulty in doing this arises from the great size of the distances between the planets as compared with all terrestrial distances. Distances on the earth are measured by surveyors by a process of triangulation. (Figure 25 (a)). First a base is measured by means of special tapes, and then, from the ends of this base line, angles are measured to some distant object. The ends of the base line and the distant object form the vertices of a triangle and once its angles are known, the lengths of the sides can be found by trigonometrical calculation. This method serves well enough for triangles with sides of a few tens of miles. For larger distances the base angles must be measured more and more accurately, for the two long sides of the triangle become more and more nearly parallel, and the whole accuracy of determination depends on measuring the divergence of these sides from parallelism. Using the diameter of the earth as a base line, or more exactly, by taking observations from two observatories, one in the northern hemisphere and one in the southern, the distance of the moon from the earth (about 240,000 miles) may be measured (Figure 25 (c)), but the distance of the sun is too large for the same method to be applied in that case. In addition the brilliance of the sun and the slight indefiniteness of the edge of its disc makes it impossible to measure the position of the edge of the sun's disc with the necessary high degree of accuracy. What we need is an obliging small planet which comes within a few million miles of the earth, and which appears only as a pin point of light to serve as a very definite mark to aim at. Then its distance in miles could be measured with precision and used to fix the scale of the whole solar system. Luckily there is such a planet.

The Minor Planets (Asteroids)

Between the orbits of Mars and Jupiter lie the orbits of a swarm of very small planets, the minor planets. Some 1500 of these bodies are known, their size ranging down from diame ters of a few tens of miles to the smallest detectable. The largest, Ceres, has a diameter of 480 miles; only half a dozen are more than a hundred miles in diameter. The smallest are probably no more than large lumps of rock. Current ideas suggest that these minor planets may be the debris of a planet of normal size which for some reason or other in the dim past was shattered to pieces. These minor planets are relatively vulnerable to the influences of the larger planets, and should a close approach occur, the orbit of a minor planet would be liable to be greatly changed by the gravitation of the large planet. The result is that the minor planet orbits are of all sorts of shapes, and show a great variety of inclinations to the ecliptic. One of the smaller minor planets, Eros, is of particular interest, since at intervals of approximately 30 years it passes within about 15 million miles of the earth, and provides an opportunity for a determination of its distance in miles. The method adopted is that a number of observatories in the northern and southern hemispheres photograph the planet on a number of nights during its close approach. The stars in these fields have previously had their positions very accurately determined, and the orbit of Eros has previously been very accurately calculated. Then from measures of the photographic plates the positions of Eros can be very accurately determined; by comparing the results from the different observatories the slight shift in position due to the different locations of the various observatories can be sifted out. The reductions and measurements are extremely complex, but they can be successfully carried out, and the value for the astronomical unit already quoted is that determined at the approach of 1931 when the work was organised by Sir H. Spencer Jones, formerly Astronomer Royal, then H.M. Astronomer at the Cape. (Figure 25 (d)).

Some Planetary Notes

Mercury is always close to the sun and so only visible just before the dawn or just after sunrise. It is a body of some 3,000 miles in diameter with no atmosphere and no satellite. It is believed to rotate on its axis in the same time as its orbital period round the sun so that it always presents the same face to the sun. The illuminated face must be at a very high temperature and the dark face very cold.

Venus is recognisable by its great brilliance and its characteristic pearly white colour. This is probably due to the reflection of sunlight from the top of an opaque cloudy atmosphere. It is not known with certainty what the rate of axial of rotation of Venus is. Various values have been quoted and there is some support for the idea that it always presents the same face to the sun as the moon does to the earth, that is, that the period of rotation on its axis is the same as its orbital period. Venus has no satellite.

The Moon

The earth is the first of the planets having a satellite. Our moon, an inconsiderable body having a diameter of only two thousand miles and a mass only one eightieth of that of the earth, presents so striking an appearance only because of its nearness to the earth. The average distance of the moon from the earth is about 240,000 miles. The moon always presents the same face to the earth so that the far side has never been seen. However, the phenomenon known as *libration* allows us to see slightly more than 50 per cent of the surface of the moon. The origin of this phenomenon is as follows: the rate of rotation of the moon on its axis is uniform but the rate of orbital movement of the moon round the earth is not uniform since the moon moves in an elliptical orbit. Because of this inequality the axial rotation of the moon is sometimes slightly ahead and sometimes slightly behind the rate of orbital turning. Sometimes a little extra strip of the eastern edge of the moon thus becomes visible and sometimes a little extra strip of the western edge. About 55 per cent of the moon's surface thus becomes visible at one time or another.

The phases of the moon are due to the varying relative positions of the earth and moon, as the moon appears to move round the earth. At new moon the moon is almost directly in line with the sun and on the same side as it. At full moon the moon is opposite to the sun. The alignment is seldom exact, but when it is the shadow of the moon falls on the earth, or that of the earth on the moon, producing an eclipse. Just after, or just before new moon when the moon appears as a crescent, the dark part of its surface may often be seen faintly illuminated (" the new moon with the old moon in his arms "). This is due to what is called earth shine, the reflection of light from the surface of the earth on to the shadowed part of the moon.

The gravitational attraction of the moon combined with that of the sun is responsible for tides in the oceans. When the moon and sun are nearly in the same straight line the gravitational effects of the two bodies combine and the tide-raising force is large. When the moon and sun are at right angles the effects tend to cancel out and the tide-raising force is small. This difference is responsible for the phenomenon of spring tides and neaps. This result may seen a trifle odd, since it is not immediately obvious that forces at right angles to one another should cancel each other out. It must be remembered that, in the case of the sun, the overall effect of gravity has already, as it were, been used up in keeping the earth in its orbit. However, gravity on the near side of the earth is slightly greater and on the far side of the earth, slightly less, than the mean. The effect of the solar gravity alone acting on the deformable ocean (and to a much less extent on the not perfectly rigid material of the earth itself), is to raise two waves on opposite side of the earth, not only immediately below the sun, but also on the remote side of the earth as well. When the gravitational force of the moon is acting along the same line this effect is reinforced and spring tides result. When the moon and the sun are in directions at right angles to one another, the crest of the two waves due to the sun lie in the troughs of the two waves due to the moon, and neap tides result.

This argument refers only to the tide-raising forces. The actual heights of tides and the water movements observed are much affected by the configuration of coasts. For example, the Bay of Fundy and the Bristol Channel, inlets which narrow from wide mouths to almost enclosed bays, produce tides of phenomenal heights, while, if the channel narrows still further, tidal bores, such as that on the Severn may result. Again, in almost enclosed bays, the rhythm of the tides may approximate to the natural period of oscillation of the enclosed mass of water and the damaging phenomenon known as range may occur. In the open oceans water movements are still not those corresponding exactly to the tide-raising forces of the sun and moon. Even in the North Atlantic the water movements may correspond more to those of water in an enclosed basin, in which the tidal wave raised moves round and round the bounding coasts with practically no movement at the centre of the area.

The surface of the moon shows a most varied topography, the chief feature of which is the large number of ring mountains, of great height and diameter. Controversy still rages, and perhaps will rage indefinitely, over the origin of these ring mountains. One school favours a volcanic origin. Another believes that they result from the bombardment of the moon's surface by meteors. The best argument for the latter hypothesis is the existence on the earth of craters produced by meteors, notably those at Canyon Diabolo, Arizona and Chubb Crater, Ungava, Canada. The number of craters certainly due to meteors is small, and all are of small dimensions compared with the large lunar craters. Any hypothesis of meteor bombardment of the moon should also involve similar bombardment of the earth, and, even allowing for the protection afforded by the earth's atmosphere and for weathering on the earth, one would expect craters of this kind to be more common than they are. The moon has no atmosphere: stars occulted by the moon are extinguished instantaneously, and theory shows that the moon's gravity is too feeble to prevent the escape into space of all gas molecules at the temperatures which prevail there.

Heights of lunar mountains can readily be calculated from the lengths of the shadows which they cast when illuminated by a low sun. The mountains turn out to be very high, and can remain so owing to the absence of wind and rain to weather them down. In all probability, the surface of the moon, which consists of some substance having about the reflective power of volcanic ash, is covered with fine dust produced by the gradual disintegration of the rocks, and perhaps the mountains themselves are more like sand dunes of fine powder than the mountains on the earth.

The smooth dark areas of the moon's surface were originally called " maria " or seas because of their appearance, but the name is a misnomer since there is no water on the moon. These areas are those visible with the naked eye as forming the " face " of " the man in the moon".

Notes on the Superior Planets

Mars, the planet next outwards from the sun is rather small: its surface is reddish in colour and in quite a small telescope the two areas at the north and south poles which are covered with hoar frost, can be seen as white spots. The axis of rotation of Mars is tilted just as the axis of the earth is tilted, so that Mars has seasons, during which the size of these frozen polar caps changes. Mars has an atmosphere, but it is very thin. It is therefore not very highly reflecting as that of Venus is, and the actual surface of the planet can be seen. There is a fairly well-known geography of the Martian surface consisting in the description of the various dark areas on its surface. The nature of these is unknown. As for the fabulous Martian canals, confidence in them suffers a severe jolt when the casual observer gets a chance to see Mars in a telescope.

How anyone can exactly delineate the large dark surface areas, let alone these alleged fine lines joining them, seems beyond comprehension, when the disc is so tiny, and when what can be seen is jumping about to a greater or less degree as the atmosphere of the earth upsets the light rays coming through it. On the question of the existence of the Martian canals opinions fluctuate a good deal. A jury of astronomers would certainly bring in the Scottish verdict of " not proven " with perhaps a small dissenting minority. Mars has two moons not visible to the naked eye. One of these is unique in the solar system in that it actually circles the planet faster than Mars rotates on its axis. As seen from the surface of Mars therefore, it will appear to rise in the west and set in the east. Considerable publicity was given at the time to the close approaches of Mars to the earth in 1954 and 1956. These phenomena occurred in the following way. The synodic period of Mars, (p. 69) is 780 days. The configuration of the sun, earth and a superior planet such that all three are in line with the earth in the middle is called an *opposition*, and, clearly, the earth is then at its closest to the planet, and the planet crosses the meridian at midnight. Oppositions of Mars thus recur at intervals of 780 days, or two terrestrial years plus fifty days. If the synodic period were exactly two years, oppositions would always occur on the same date with the earth always at the same point in its orbit, and Mars at the same point in the sky. The surplus fifty days cause the time of opposition to progress round the year and the position of opposition to advance round the orbits. The orbit of the earth is closely circular: that of Mars distinctly elliptical. The two orbits are so placed that the distance between them has a minimum value of 37 million miles near a part of the earth's orbit which we traverse at the end of August, and a maximum value of 62 million miles for the part of the earth's orbit traversed in February. Thus oppositions of Mars occurring in August and September are much closer than February and March oppositions. In 1954 Mars was in opposition on June 25, and in 1956 on September 12, the latter very close to the most favourable date possible. Later oppositions will occur at dates delayed in each opposition year by 50 days, and close oppositions will recur only when opposition date has advanced once more to the favourable season, i.e. about seven oppositions later than 1956 (7 × 50 days is close to one year), that is in 1971.

Next come the minor planets, one of which, Eros, has already been mentioned. The largest of them are Ceres, Pallas, Vesta and Juno. An interesting new minor planet discovered in 1950 is Icarus, distinguished by the fact that it has a very elongated orbit carrying it nearer to the sun at perihelion than any other planetary body.

Jupiter, whose orbit is next, is the largest of all the planets. It is a globe with diameters of 88,700 miles and 82,800 miles, these being the equatorial diameters and polar diameters respectively. This high degree of flattening indicates that the visible surface of Jupiter can hardly be solid, and that what is seen is the upper surface of the atmosphere, flattened by the rapid rotation of the planet on its axis which occupies a period of only 9h 50m. This high degree of flattening indicates that the visible surface of Jupiter can hardly be solid and that what is seen is the upper surface of an atmosphere distorted by the rapid rotation of the planet on its axis. Surface detail, which enables the rotation to be studied, gives a rotation period of 9h 50m near the equator, and 9h 55m in higher latitudes. The surface features can be glimpsed with only the slightest of optical aid. Well-marked belts of a red colour, and showing fine detail, encircle the planet to north and south of its equator, much as the trade wind belts do on the earth. Other belts lie further north and south. On the south equatorial belt is an indentation which at one time marked one of the most striking features of the surface of the planet—the area known as the great red spot. This has now faded and only the indentation in the belt which it once occupied remains visible.

Jupiter is a cold place: estimates of its temperature depending on the measurement of the infra-red radiation received from the planet put its temperature far below freezing-point, and current ideas suggest that the outer atmosphere, which it is known contains methane and ammonia, resembles nothing so much as the cooling fluid in an ammonia refrigeration plant. It is presumed that the planet has a rocky or perhaps ice-sheathed core a good deal smaller than the visible disc presented to observation. Jupiter, more than most of the planets, deserves mention in a book designed for those who wish to teach themselves astronomy, since it is possible with the aid of no more than a good pair of field glasses supported against some rigid structure, to get a fair glimpse of its surface. Even more striking are the four brightest of the planet's twelve moons. The remaining eight require very powerful astronomical equipment for their observation, but the four brightest are readily seen and their movements from hour to hour can be readily followed. In particular it is very often possible to follow the movement of one of the moons as it passes behind the planet, or is eclipsed in Jupiter's shadow, or to see the moon itself and its shadow passing in front of the planet's disc.

Saturn, the next planet, is slightly smaller than Jupiter with a diameter of about 70,000 miles. The globe of the planet is even more flattened by rapid rotation than is that of Jupiter. Little or no surface detail can be seen even in a quite powerful telescope, and it is assumed that this surface is also gaseous, and at a very low temperature. It is possible to estimate the mass of any planet possessing a moon in terms of the sun's mass. Roughly speaking this is done by a calculation which compares the gravitational force necessary to keep the planet in its orbit round the sun, with that necessary to keep the moon moving round the planet. This type of argument applied to Saturn gives a very low value for the mass, and putting this against the total volume of the planet it is possible to deduce the very astonishing result that the density of matter in Saturn is less than that of water. This represents the average volume, and, assuming that the centre of the planet must be formed of dense rock we are forced to the conclusion that, to balance matters out, most of the planet must be of very tenuous, probably gaseous, material. The chief glory of Saturn is, of course, its ring system which looks like a disc of material, thin and flat as if it were cut out of cardboard, surrounding the equator of the planet. The constitution of these rings formed, at one time, one of the capital problems of astronomy, and the problem was solved theoretically by Clerk Maxwell many years before the correctness of his result was confirmed observationally. The rings consist of a swarm of tiny moons each possibly no more than a few thousandths of an inch or a few inches—the exact size it is impossible to deduce —in diameter. At all events the rings are not opaque for when Saturn in its orbital motion passes in front of a star the star is not extinguished but only dimmed.

Present ideas of the origin of the rings are that a satellite of Saturn once approached to a certain critical distance from the parent planet, at which, as can be demonstrated theoretically, the internal stresses in the satellite would be sufficient to shatter it to fragments. The particles forming the rings are thus presumed to be the remnants of a moon of Saturn. In addition Saturn has nine satellites of which the majority are very faint.

The planets already mentioned have a proper place in the astronomy which anyone can teach themselves, although in fact, in describing them, we have gone beyond our strict mandate. The remaining planets are all telescopic objects, although the first, Uranus, lies only just below the limit of visibility, and might just be glimpsed with the naked eye as an extremely inconspicuous object under the most favourable conditions if one knew where to look. It was discovered accidentally by Herschel who at first wished to give it the name (fortunately it did not find favour) of Georgium Sidus—George's star, after George III. It has a diameter of about 30,000 miles and has five satellites whose orbits are all in a plane (just as are the orbits of the planets round the sun), but this plane is amost at right angles to the ecliptic. Instead therefore of seeing the satellites shuttling back and fore with respect to the planet, we can sometimes see the orbits of the satellites of Uranus almost flat on, as circles centred on the planet. Still further out comes the planet Neptune, discovered in 1846 as a result of independent mathematical predictions based on irregularities in the motions of Uranus, by Adams and Leverrier. The diameter of this planet is 28,000 miles and it has two satellites Triton and Nereid, the latter discovered in 1950.

Lastly, in 1931 Tombaugh discovered Pluto of which little is known except the details of its orbit. The diameter is 3,600 miles according to a recent estimate and no satellite has yet been discovered.

Perennial Questions

Two questions, both unanswerable with certainty, are always asked about the solar system. One is whether there is life on any of the other planets. The other is how the system came into being and whether or not it is unique.

The existence of life as we know it depends on the presence of moderate temperature conditions in which large molecules such as those of the proteins can persist, and on the presence of water vapour and oxygen. It will immediately be apparent that no life as we know it can persist without special equipment on planets like Jupiter, Saturn and the other exterior planets, where the temperature is low, and where the most abundant gases present are those which are noxious to almost all forms of terrestrial life. Equally no terrestrial life could exist on a planet like Mercury where there is no atmosphere and no water vapour, and where on the illuminated side the temperature may far exceed that (about 50 degrees centigrade) which produces burns and tissue destruction. The only remaining candidates in the solar system are Venus and Mars. Mars, in spite of being the favourite of fiction, probably does not present very favourable conditions for the maintenance of life as we know it. The temperatures, though less extreme than on the other planets, are rather low; the atmosphere is very thin: carbon dioxide is present, water vapour possibly present, and oxygen probably absent. Conceivably Venus may offer a more favourable abode for life, but because her atmosphere is opaque we know little or nothing of conditions on the solid surface of the planet. The atmosphere contains abundant carbon dioxide, a gas, which, in natural occurrence on the earth, is often associated with plant life. Water vapour and oxygen have not been detected. The fundamental difficulty of detecting water vapour and oxygen in the atmosphere of a planet arises from the abundance of these same substances in our own atmosphere. Generally speaking the presence of these vapours in the atmosphere of a planet could only be detected through a very slight intensification of certain effects which are already demonstrating the obvious presence in our own atmosphere of large quantities of both these substances.

Theories of the origin of the solar system are legion: most of them have started from a consideration of the curious regularities exhibited by the various planets: the small planets are nearest to the sun: the giants are in the middle, while the outermost are again small. Those near the middle of the series have the most numerous satellites ; there is also a rule called Bode's law which, roughly, but only very roughly, enables one to write down the distances of the various planets from the sun by means of a simple arithmetical formula. Some, but not all, astronomers regard this as a mere coincidence.

The question of the origin of the solar system has been the subject of numerous theories, and heated and inconclusive debate. Since no one witnessed the original events, and since they cannot be repeated, they will probably remain a subject for controversy indefinitely. A fact, possibly of great significance, is that the age of the earth and the age of the Universe are of the same order of magnitude, and this might imply that conditions necessary for planet formation can only have obtained during the early stages of the history of the Universe. Numerous theories which depended on encounters between our sun and another star have now been swept away, and the theory in vogue at the present time relies on the idea of a mass of gas within which eddies formed, and provided the conditions which enabled the solid planets to be condensed.

CHAPTER IV

THE SUN

Solar Energy

The sun is the source of all the natural light and heat of the earth, apart from a certain internal production of heat by radio-activity. It is the source of all the combustible fuels available on the earth. Coal is derived from prehistoric forests which drew their energy for growth from the sun in ancient times. Oil and petroleum, according to some views, are of animal origin, owing their growth to the sun. Wind power, used for pumping water and generating electricity in some parts of the world, represents solar energy absorbed by the air and water of the atmosphere and oceans, giving up this energy in the form of the motion of masses of air. Water power resources represent a conversion of solar energy first into the stored or " potential " form of evaporation energy of water, and then into the potential energy possessed by water lifted against gravity from sea level to a reservoir. All these forms of stored energy are concentrated: but only part of the solar energy available has undergone this concentration: the rest, by a fundamental law of physics, has become less concentrated and less available. It may even be claimed that the sun is the ultimate source of the fissile material used in the industrial generation of atomic power, for it seems probable that these heavy elements were originally synthesised in the interiors of stars.

Energy received from the sun directly is such that on each square centimetre there falls in each minute about two calories of heat energy, an amount sufficient to raise the temperature of two grammes of water by one degree centigrade. It is rather less than the energy received from an electric fire at a distance of a few feet. The total energy falling on a large area of land is enormous, but the difficulty is to concentrate it and make it available in useable form. The only parts of the world where the direct use of solar energy is at all likely to be practicable are those regions of very low rainfall where the sun shines all or nearly all the daylight hours: it is, there-fore, almost inevitable that these regions should be desert areas, sparsely inhabited, where no use or only very slight use for the collected energy can be found. It is not surprising therefore, that although small scale solar energy plants have proved practicable, the only attempts to utilise solar energy commercially have been unsuccessful.

Without a telescope, preferably of very special design, there is little that can be done in the way of solar observation; but the sun is the nearest example to us of a star, and it is the only star, which appears in any instrument yet constructed as more than a mere point of light. We shall therefore again go rather outside our brief of "Teach yourself astronomy" to say something about the structure and properties of the sun.

Solar Structure

The sun has a diameter of 864,000 miles: it is a globe of gas, mainly hydrogen, having a temperature at the surface of about 6,000 degrees centigrade. The temperature and density of the solar material must, so all theoretical arguments indicate, rise steadily the deeper below the surface one goes. At the centre of the sun the material is still gaseous, but is at an enormous density and pressure. The temperature there must be of the order of 20 million degrees which is sufficiently high to cause splitting and combinations of the nuclei of atoms. These processes produce atomic energy in the form of radiation which filters very slowly out through hundreds of thousands of miles of solar material, being absorbed and re-emitted many times, heating up, or rather maintaining the heat of the material through which it passes: finally this energy, much changed by its lengthy journey, arrives at the surface and goes out into space as light and heat waves. This is believed to be the mode of generation and emission of energy in all stars: it was the first suggested example of atomic energy, proposed a quarter of a century ago, as being the only conceivable way in which the stars could produce and continue to produce their tremendous energy output for hundreds of millions of years.

The picture sketched here is a very compressed outline of the present knowledge of the stars, based on a vast mass of experimental and observational knowledge and on theoretical deductions made from it. There is no space to go into details here, but it can be said that every piece of information used in building up this picture of the sun has a very sound basis in experiments carried out in terrestrial laboratories; much of this information has a very immediate practical application to industrial and other processes.

Physics Applied to the Sun

One example of the kind of arguments used, leading to

results which can be verified by observation, must suffice. The laws describing the way in which hot bodies radiate have been discovered in the laboratory, and have been fitted perfectly into the structure of modern physical theory. One of these laws says that as the temperature of a hot body increases, the total energy radiated per second from each square centimetre of its surface increases in a certain way. Now, imagine a region just underneath the surface of the sun: the pressure of the gas in this region must be sufficient to support the weight of all the layers of gas lying above this level; this weight is due to the gravitational attraction of

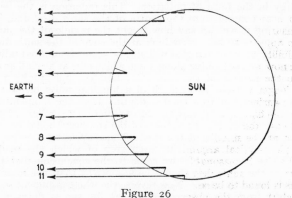

Figure 26

Rays 1, 2, 3, 9, 10, 11 *originate nearer the solar surface than rays* 4–8.

all the rest of the sun. In the same way the gas pressure in the atmosphere of the earth at sea level must be sufficient to support all the air which lies above. In the atmosphere of the earth we know that the pressure decreases as we ascend, since there is less and less air above to be supported. In the same way, in the sun, which is all gas—nothing but atmosphere as it were—the pressure must increase as we go inwards from the surface. This can be shown to imply that the temperature must also increase (which is also true as we descend through the atmosphere of the earth). But now, if the temperature increases, by what we just said about the laws of radiation, the amount of light and heat radiated by each square centimetre of the surface of each layer of the sun's atmosphere must also increase. Since the sun is made of gas, it is partly transparent, that is, when we look at the sun we shall see radiation coming not only from the very surface but also from a certain distance below the surface. Let us say, for the sake of argument, that this distance is a hundred miles: that is we can think of the sun as being made of hot gas, partly transparent, like a sort of hot fog, in which visibility is a hundred miles. Now let us, in imagination, look at the middle of the sun's disc. (Figure 26.) Then our range of visibility of 100 miles takes us in to a position 100 miles below the surface where the temperature has a certain value, and correspondingly where matter emits a certain amount of energy in the form of radiation. This radiation is the light and heat coming from the centre of the sun's disc, and, by measuring it we can say how bright the middle of the sun's disc appears to be. Now look at the edge of the sun's disc. Visibility in solar material will still be about a hundred miles, but now we are looking along a line which cuts at a small angle into the surface of the sun. Our hundred mile probe now no longer reaches to a depth of a hundred miles below the solar surface but to a smaller depth. Here the temperature is not so high, and the outpouring of radiation is smaller. That is to say the surface brightness of the sun here should be less than at the middle of the disc. Thus, by using a series of simple physical arguments, each step of which the reader will agree is reasonable, we come to the conclusion that the edge of the sun's disc should be less bright than the middle. This is found to be so. Now reverse the whole argument so as to start from the observed fact that the sun is dimmer at the edge than the middle. Then, by actually measuring the way in which the brightness changes, and by refining the physical arguments sketched above, we can determine the way in which the temperature changes as we go inwards from the surface of the sun. This kind of reasoning does not provide us with information about layers deep below the solar surfaces (probably not more than 100 miles or so), but it does give us a great deal of information about the most important part of the sun for us, the outer skin which we actually see.

Although the sun is very remote, it is thus possible to begin to apply knowledge of physics gained in the laboratory to the study of the sun's structure and behaviour, and to make deductions from these laws which can be checked by observation. The picture of solar structure which we have at the present time is a jig-saw puzzle, some of whose pieces are missing, while, to continue the metaphor, others which we have in our possession probably belong to some completely different puzzle. We have enough to be fairly sure of the general outlines of a good deal of the picture, but many important details are missing, or distorted, and there is much controversy over these doubtful parts of the picture. Unfortunately, as we have said, almost the whole of this picture lies well outside the terms of the brief " Teach yourself astronomy ". The physical structure of the sun cannot be investigated at all without instruments, and therefore should not command attention in this book. But, on the other hand, the sun is the only star for which a fairly complete observational investigation is ever likely to be possible. We cannot ignore it if we wish to give a balanced picture of the way in which astronomy is now developing. What we have dealt with so far—the elements of the astronomy of position and of navigation, the constellations, the movements of the planets, and so forth, are all important departments of astronomy, but they are departments in which the main bases of knowledge are all firmly founded. What can happen in these departments in the future is no more than a slight improvement of the superstructure of our knowledge. It is in astrophysics, the knowledge of the physical structure of the stars and of the sun, that the most exciting future developments lie. We cannot therefore omit the sun from our discussion, and we shall compromise by giving a brief outline of out present solar knowledge.

We have already outlined the way in which a typical problem of solar structure may be approached. This problem is, in effect, a problem of the solar surface, that part of the sun which is accessible to direct observation with the eye or the photographic plate. The temperature of the solar surface has been determined by comparing the quality of the light which it emits, that is the proportion of energy emitted in light of various colours, with the light emitted by materials in the physical laboratory which are maintained at known temperatures. The results which have been found by this and other methods yield values close to 6,000 degrees centigrade. At this temperature, which is almost twice as high as that of any terrestrial furnace, all matter is gaseous, and almost all chemical compounds are dissociated into their elementary atoms. Physically speaking, however, it is not a very high temperature, and is far below that needed to disturb the nuclei of atoms: or, to put the matter another way, it is far lower than the temperature generated in regions where splitting of atomic nuclei is taking place   The temperature which obtains for a brief fraction of a second in the material of an exploding atomic bomb, before the material begins to disperse and cool off, is much more reminiscent of the temperature which is continuously maintained at the centre of the sun, rather than at its relatively cool surface.

As we have seen, when we look at the sun we are apprehending not a simple surface but something which has a structure in depth. This outer layer which we can observe directly is often called the solar atmosphere by scientists. Like the atmosphere of the earth it has changes—weather if you like—even though it is not, as the earth's atmosphere is, bounded below by a solid surface. On the contrary, in the sun all is gas, and the solar atmosphere merges imperceptibly into the deeper layers of the sun's structure.

Sunspots

The most striking of these climatic phenomena are the sunspots, areas of relative darkness, and lower temperature, which, however, seem dark only by contrast with the blinding light of the rest of the solar surface. Sunspots entered modern astronomy through telescopic observations, but they were known to ancient astronomers since, rather rarely, very large spots are visible to the naked eye under suitable conditions, such as, for instance, when the sun is seen through a thin layer of cloud. Sunspots occur almost at random, but there are certain regularities about the times and positions of their appearance. The sun is rotating about an axis, not as a solid body must, with each portion taking the same time for a revolution, but in the manner characteristic of a sphere of gas, in which the surface near the equator revolves in a shorter time than that near the poles. The rotation period of the solar surface near the equator is about 25 days, rising to about 34 days near the poles. The same point on the sun's equator is opposite to the earth about every 27 days. This is the synodic period of the axial rotation of the sun, and the orbital motion of the earth. A sunspot which lasts long enough may thus reappear at intervals of 27 days. Most sunspots last only a few days at most before gradually disappearing, but some large groups are sufficiently long lived to make several reappearances. The form and number of the spots in the group will usually change from appearance to appearance.

It will be noted that the fact that the sun rotates enables us naturally to define and use the ideas of a north and south pole on the sun, of a solar equator half-way between, and to

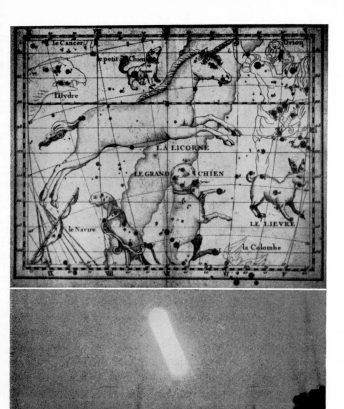

PLATE I.

Top : Celestial congestion in the 18th century. Two pages from a French translation, dated 1776, of Flamsteed's Celestial Atlas, showing Orion, Lepus, Columba, Monoceros, Canis Major and Minor, and parts of Cancer, Hydra, and Argo.

Bottom : Time exposure showing full moon rising in south latitude 26°. The angle of rising is far steeper than in higher latitudes.

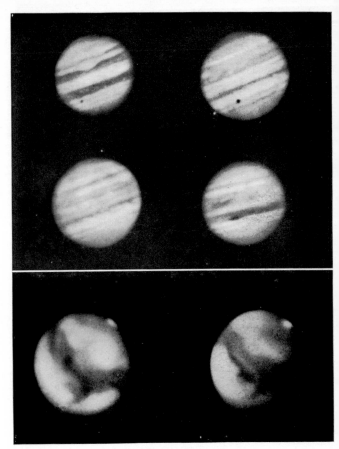

PLATE II.

Top : Four views of Jupiter, with, in one, the satellite Ganymede and its shadow.

Bottom : Two views of Mars, showing polar caps, and surface detail. At right Mars is seen gibbous.

(*Mount Wilson Observatory photographs.*)

PLATE IV.

Top : The Southern Cross and Coal Sack. The photographic plate exaggerates the brightness of blue stars as compared with red.

(*Harvard Observatory photograph.*)

Bottom : Kappa Crucis (Herschel's Jewel Box), a cluster of stars of a wide range of colours.

(*Radcliffe Observatory photograph.*)

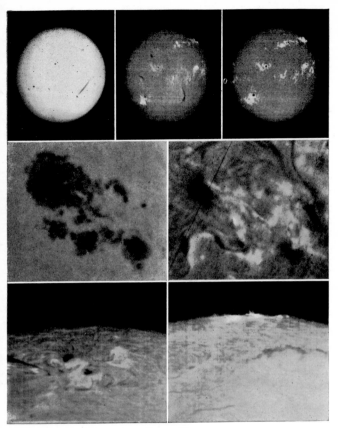

PLATE V.

Solar photographs. Three views at top are in ordinary light and in selected colours. Sunspots and associated bright clouds are seen.

Middle : Direct photograph of sunspot group, and of same group in selected colour.

Bottom : A sunspot group nearing the edge of the disc as the sun rotates. The photographs (made in a selected colour) are separated by an interval of two days.

(McMath-Hulbert Observatory photographs.)

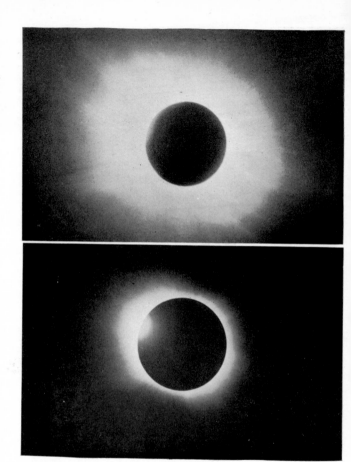

PLATE VI.

The total solar eclipse of October 1st, 1940. Top: total phase with corona visible.

Bottom: The end of totality with light shining down a lunar valley (diamond ring effect).

(Cape Observatory photographs.)

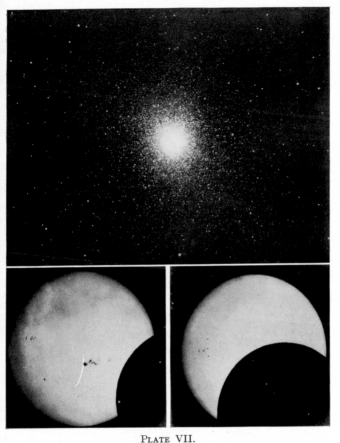

PLATE VII.

Top : The Globular Cluster in Hercules.

(*Mount Palomar photograph with 200-inch telescope.*)

Bottom : Partial eclipse of sun of November 23rd, 1946. An interval of one hour separates the two exposures.

(*Associated Press Photo : by U.S. Naval Observatory.*)

PLATE VIII.

Top : Making the mirror for the 82-inch telescope of the McDonald Observatory, Texas.

(Photograph, Warner and Swasey Company, Cleveland, Ohio.)

Bottom : Foucault test of a bad mirror. All minute surface scratches readily show up.

think of the solar latitude of any point on the sun's surface. Sunspots occur in greatest numbers just north and south of the solar equator. (Figure 27 (a)). They also vary in their incidence with time. Although sunspots of almost any size and in almost any numbers may occur at any time, statistics show that on the average they appear in greatest numbers every eleven years (Figure 27) (c). During the early part of each

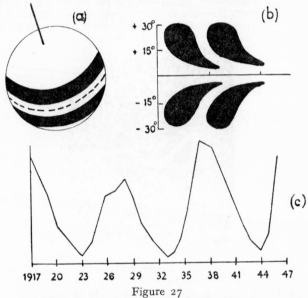

Figure 27

(a) Latitude zones in which spots occur; (b) spots occur in positions corresponding to the black areas, i.e. in lower latitudes later in each spot cycle; (c) graph showing numbers of spots observed.

11-year cycle a few rather small spots appear in relatively high northern and southern latitudes. As the sunspot cycle wears on, the spots become more numerous and larger, and appear nearer the solar equator. Finally, at the end of a cycle, the last spots of the old cycle are appearing close to the equator simultaneously with high latitude spots of the new cycle.

Exactly how the sunspot cycle is generated it is impossible to say. It must reflect some deep-seated change in the condition of the sun. The spots themselves are in the nature of

D

vortex structures ranging in size from a few thousands of miles in diameter up to spot groups with sizes of several hundred thousand miles. They are the seat of complex rotary and vertical motions of gases, and often of very intense magnetic fields, which, at the blackest part of each spot (the *umbra*) may exceed the largest magnetic intensities produced in terrestrial laboratories. The spots shade off into a zone of intermediate brightness (the *penumbra*) where there are often many complex radial tongues of gas. Round the spots may appear streaks of extra brightness while over them,

Figure 28

Sketch of sunspot pair, showing granulation of surface and prominences over the spots.

stretching perhaps hundreds of thousands of miles out into space, there may be areas and wisps of gas. These last structures are called *prominences* and require special equipment and conditions for their observation. Spots often occur in pairs of opposite magnetic polarity—like the two opposite magnetic centres at the ends of a bar magnet. In some cases no second spot can be observed visually but the magnetism can be detected because of certain rather subtle effects which it has on the light emitted from this region. (Figure 28.)

It is possible that the sun emits rather more light and heat at times of sunspot maximum: this rather paradoxical result would be due to the fact that although the area occupied by a sunspot itself is darker and therefore emits less light, the surrounding area of the sun's surface will be more disturbed and the extra bright streaks already mentioned may more than compensate the loss by their increased radiation.

Terrestrial Effects

One of the effects of the sun on the earth is produced by the absorption of the invisible ultraviolet light of the sun in the atmosphere of the earth, particularly by water vapour and ozone. This is fortunate since the human skin is sensitive to ultraviolet light which may cause burning or tanning or even much more serious damage. At high altitudes the protective cover is diminished because of the reduced thickness of the air barrier: in very dry desert climates water vapour is almost absent and so there is less protection. It is not unknown in some places for the very intense ultraviolet radiation received on the skin day after day over many years to produce cancerous lesions. We have therefore good reason to be thankful for our protective air blanket which merely lets through enough of the ultraviolet light to give us a good comfortable tan on our seaside holiday. On the other hand it does mean that in ordinary observation of the sun we are, as it were, colour blind because a great part of the light emitted by the sun never reaches us. However, in recent years a good deal of work has been done by projecting rockets carrying suitable automatic recording instruments to heights well above the main part of the atmosphere of the earth: our knowledge of this most important part of the energy emitted by the sun has in consequence rapidly increased.

The energy absorbed by the earth's atmosphere does not, however, disappear. The upper layers of the atmosphere of the earth are changed by the act of absorption. In these very tenuous layers at heights of 60 miles or so above the earth, the effect of absorption of ultraviolet light is to cause the emission of electrons from the atoms of the air. An atom of which an electron has been emitted in this way is called an *ion* and the production of an ion is called ionisation. At normal pressures, emitted electrons are snapped up by wandering ions almost at once, but at very low pressures there can be a perceptible interval between the emission of an electron and its recapture. This means that at high altitudes the earth's atmosphere consists of matter in a partly dissociated state. When an electron is recaptured by an ion, the energy originally required to split the electron off an atom is re-emitted as radiation. However, although the original energy causing the ionisation came in a beam from the sun it is re-emitted in all directions giving a generalised diffuse faint light. (Figure 29.) At night these recombinations are still going on and are responsible for the faint light of the night sky.

The existence of these electrified layers at great heights is

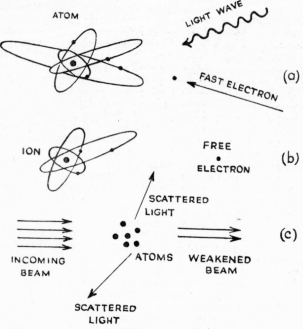

Figure 29

(a) An atom is ionised by a light wave or fast electron and is split up as at (b); (c) shows how a beam of light is scattered by this process.

of importance for many terrestrial purposes. Wireless waves entering them are turned from their original direction much as light waves incident on a layer of heated air may be turned so as to produce a mirage. This turning enables the radio waves to be received beyond the visible horizon, and indeed, by repeated reflections between the upper layers and the surface of the earth, they can reach right round the circumference of the globe. Wireless reception of London in Australia or of

New York in Britain is entirely due to this repeated reflection of the waves.

The behaviour of the waves depends on their wavelength. The reflecting layers have a complex structure and reflections of waves of different wavelengths or frequencies takes place at different heights. Since the structure of these atmospheric layers depends on the emission of ultraviolet light from the sun, it follows that if the intensity of this light should change, then it may be necessary to switch a long distance radio transmitter from one wavelength to another. As the sun goes through its cycle, it is believed that the average total intensity of ultraviolet radiation from its surface changes, and there is a well-marked correlation between the spottedness of the sun and the most suitable frequencies to be used for long-distance radio communication on the shorter wavelengths.

Solar Flares

However, in addition to this general variation, in rhythm with the sunspot cycle of eleven years, there are far more catastrophic short period changes. Of recent years a great deal of attention has been paid to what are called *solar flares*. Only most exceptionally are these large enough and bright enough to be visible without the aid of special instruments. In brief what happens is that in the neighbourhood of a sunspot there will suddenly appear a small area (that is small by the standards of solar dimensions) of intense brightness, which after reaching a peak of brilliance, fades away in a time of the order of twenty minutes, perhaps changing its shape and becoming drawn out into a filament as it does so. Just at the instant when the flare appears, sensitive magnetic instruments on the earth will register a small disturbance. At the same time, a very intense outburst of ultraviolet radiation comes from the flare region which, on being absorbed by the earth's upper atmosphere, so steps up the number of free electrons in it that short-wave radio waves are absorbed and communication is interrupted. This effect may last for some hours on the side of the earth turned towards the sun at the moment of the appearance of the solar flare. However, this is not all. It is almost certain that the flare area, in addition to emitting ultraviolet and visible light, also sends out a stream of positively charged particles (ions), including ions of calcium. These are sprayed out into space and if the flare happens to be suitably placed on the sun, then by the time the particles have covered the distance from the earth to the sun, the earth will run into the stream of ions. When this happens the magnetism of the earth will cause them to swerve and to strike the upper atmosphere of the earth mainly in the neighbourhood of the north and south terrestrial magnetic poles. The ions then pick up free electrons and in other ways stimulate the upper layers of the earth's atmosphere, so as to produce the bright curtains of light known as the *aurora borealis* and *aurora australis*. (Figure 30.) At the same time this stream of moving electricity produces magnetic distur-

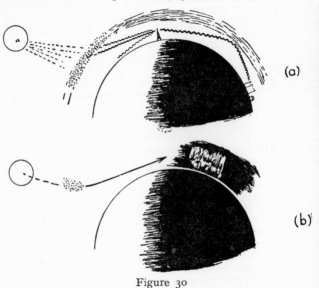

Figure 30

(a) Normal transmission of radio waves and interruption by radiation from a solar flare; (b) ions from the sun cause aurorae.

bances called magnetic storms. It will be clear since the earth and the flare region on the sun's disc have to be in suitable relative positions that it will not always follow that there will be aurorae, radio fade-outs and magnetic storms each time there is a flare. Nor will it be certain that a flare will occur even when the sun's surface is greatly disturbed by the presence of a large group of sunspots. However, on a number of occasions sunspot groups of such great size have appeared, that flares seemed almost certain to occur and successful predictions were then made.

Although it is impossible for the amateur, lacking instruments, to observe solar flares, their effects can often influence him. The radio " ham " may find communication interrupted by fading due to flares: radiograms to distant friends may be delayed: in Northern Europe or America or in New Zealand aurorae may be seen. Auroral phenomena are in fact much less rare than one might imagine, and abnormal brightness of the moonless night sky in places remote from artificial light can often be detected even when really striking aurorae cannot be seen. The remaining effects, such as the magnetic disturbances are, however, too small to be detectable except by special instruments: even in a most violent magnetic storm the disturbances are much smaller than those which can be measured even with moderately good instruments, such as, for example, those possessed by a school physics laboratory.

Figure 31

Observing the Sun

It is possible with the aid of a small telescope or *some* types of field glasses to observe sunspots of large size. DO NOT UNDER ANY CIRCUMSTANCES LOOK DIRECTLY AT THE SUN THROUGH A TELESCOPE OR FIELD GLASSES UNLESS THE INSTRUMENT IS PROVIDED WITH DARK COVER GLASSES OF ADEQUATE

STRENGTH SUPPLIED BY THE MAKERS SPECIALLY FOR SOLAR OBSERVATIONS. Clamp the telescope or glasses to a suitable post or stand—preferably you should have some way of swinging the instrument to follow the sun as the earth rotates. Point the instrument at the sun by receiving its shadow on a card: when the telescope is pointing at the sun the shadow of the tube will just be a circle. Then having done this you will find the sunlight coming through the instrument and by focusing the image on to a card you will obtain a picture of the sun and of any sunspots which may be present. A good plan is to fix the card with wires to the body of the instrument in approximately the right position. (Figure 31.) Then you will have fewer things to manage. You may have to observe for several days before big spots appear even at times of maximum spottedness, while at sunspot minimum the period may be longer. It is however worth while to persevere. You will also be able to see that the edge or limb of the sun is darker than the centre of the disc.

In professional studies of the sun, very long telescopes producing very large solar images are used. These are often in the form of fixed instruments of great length into which the sunlight is fed by a mirror slowly turned by a motor in such a way that the reflected beam of sunlight goes down the tube. Telescopes of this kind with lengths up to 150 feet, giving a solar image more than a foot in diameter have been constructed, and with their aid detailed studies of phenomena on the solar surface have been undertaken.

Radio Waves from the Sun

The sun emits not only visible and ultraviolet light; it also emits radiation of greater wavelengths : those just beyond the range of visibility we can detect readily enough by their heating effect on our skins or, more scientifically, by special photographic emulsions or sensitive heat receivers. At still greater wavelengths—of the order of a few centimetres or metres, we enter the band of ultra-short-wave radio waves used for radar. Generators and receivers of such waves were developed during the war for the first time, on a scale exceeding that of the laboratory, and apparatus of this kind has been put to a variety of uses, some of them in astronomy. For example, signals on ultra-short-wave radio which will pass through the electron layers in our upper atmosphere have been used to produce echoes from the moon. The moon emits no radiation of this kind itself, and merely served as a

reflector in these experiments.   However, in the last few years
it has been discovered that the sun, certain stars, and certain
kinds of nebulae and gas clouds in space are emitters of short
wave radio waves, which can be detected by means of ingenious
and complicated apparatus.   The last decade has seen the
growth of the science of radio astronomy, a completely new
and rapidly growing field which bids fair to rival ordinary
visual astronomy.   This is, however, no field for the amateur,
and we must here confine ourselves to a brief reference to the
radio astronomy of the sun.   The short wave radio emission
of the sun is of two types: one type represents emission by the
sun as a whole and is more or less steady; but in addition,
sunspot areas and flare areas emit during times of disturbance
and at such times the radio intensity received from the sun
increases greatly.   Reception is possible on a variety of
wavelengths and there is some uncertainty as to how the
phenomena in the different wavelengths can be fitted together
to form a coherent picture of what is going on.   One difficulty
is that, considered as telescopes, the various forms of radio
receiver which can be used are not very efficient.   In order
to form a detailed picture of what is in front of it, a telescope
or an array of aerials must be many times larger than the
wavelength of the radiation which is being employed.   For
a visual telescope employing waves of a wavelength of, say,
50 millionths of a centimetre the instrument diameter is
always very many times larger than the wavelength of the
radiation.   For a radio receiver working on, say, 50 centimetres wavelength, the aerial array, which has to be swung to
follow the sun, cannot be many times the wavelengths of
the radiation without becoming very large.   The instrument
developments of the past few years have been devoted, on
the one hand, to the improvement of resolution by the use of
intricate aerial and recording systems, and on the other, to
increasing sensitivity by increasing the size of metallic
reflectors used to concentrate the incoming rays.   The largest
equipment now in use, the 250 foot diameter reflector at
Jodrell Bank, Cheshire, is of a size undreamed of a few years
ago.   Installations of more modest size have sprung up almost
by the dozen in countries all over the world.   Radio astronomy
can be carried on in all weathers, and, when sources other than
the sun are being investigated, at any time, day or night,
that the source is above the horizon.   With improved resolution and sensitivity it is impossible to predict limits to the
potentialities of this new astronomy, which is complementary
to ordinary visual astronomy.

D*

Solar Appendages

We have already described sunspots as being disturbed areas of the solar surface and have mentioned that very often there are to be found, above and near them, arches of luminous gas, often of very great size, which are known as prominences. These are not visible by ordinary direct methods of observation, but a variety of special methods exists which can be used. Prominences are regions in which free electrons are joining up again with ions (atoms deprived of one or more electrons). It is a characteristic of ions that when they re-capture an electron they emit light in certain colours or wave-lengths which are peculiar to the particular type of ion involved (conversely, one way of producing an ion is to throw a beam of light of the right colour on to a group of the atoms; the atoms then pick out the special colour which suits them and each ejects an electron). Thus the light of prominences does not resemble the light emitted by an ordinary hot body, such as a lump of glowing iron, but is more like the light emitted by the type of street lamp which has become popular of recent years, employing either mercury vapour or sodium vapour. In each of these types of lamp the gas atoms are ionised (that is electrons are split off them) by the collisions produced by sending through the gas a stream of fast moving electrons. This stream of electrons is in fact no more than an electric current passing through the gas and is maintained by applying a high voltage between two wires inside the lamp. Then, later on, when the ions so produced strike the walls, or otherwise get possession of an electron, they recombine and emit light. In this way the energy used in maintaining the electron streams is converted into light.

To revert to the prominences; if we can manage to separate out the special wavelengths emitted by the recombining ions in prominences, the brightness of these clouds can be greatly enhanced as compared with the rest of the solar surface, which emits light of all wavelengths, and in this way the prominences can be made visible. Prominences contain a high proportion of hydrogen, and, if the hydrogen light can be separated out, the distribution of hydrogen in the prominences can be studied. A whole new subject of research has been opened up in this way; the prominences exhibit remarkable structures—arches, plumes, streamers and the like—many of which are in very rapid motion.

The only hope which the amateur has of seeing a solar prominence is by arranging to view a total solar eclipse (see Chapter VI) when the bright light of the sun's disc is cut off

by the moon and the relatively faint prominences can be seen
with the naked eye. In France and later in the U.S.A.
and elsewhere a technique has been developed in which an
artificial eclipse is created   In desert regions where the air is
free of clouds, moisture and dust, and where, in consequence
the proportion of light scattered in the atmosphere is small,
it is possible to hold up a threepenny piece (the silver kind)
at arm's length and to cover the solar disc so effectively
that one can look directly at the sun without discomfort.
This is never possible in a wet climate like that of England.
If one were to go to a high altitude where there was even
less air to cause scattering, one might attain to a condition
where the sky was relatively dark even quite close to the sun,
and then it might be possible to see the prominences peeping
round the edge of the threepenny piece.   If it is ever possible
to travel to other planets by rocket ship, prominences will
be one of the sights to look for.   One might view the sky
from behind a screen just large enough to hide the sun's disc,
and then, against the complete blackness of the sky, see the
solar prominences standing out as pink arches and loops.   That
is for the possible future : in the present there are several
stations at great altitudes in mountainous regions, including
one on the Pic du Midi de Bigorre in the Pyrenees and one at
Climax, Colorado, U.S.A., where the technique developed by
M. Bernard Lyot the French astronomer, is applied.   With
many complications and much refinement of technique the
method used is, at bottom, the simple one of putting up a
disc to hide the sun and observing a ring of sky round its
edge.   Both groups of workers have produced films of the
prominences observed in this way which show the most
delicate structure and movements of the gases.

The Chromosphere and the Corona

So far we have spoken of the solar surface and a certain depth
below it : of the sunspots which penetrate like cool funnels
some way into the surface : of the prominences which arch
over the spots and of solar flares which are mainly but not
entirely a phenomenon occurring in the surface and just above
it.   The sun has two more appendages which we have not
yet mentioned—the chromosphere and the corona.   The
chromosphere (literally, " colour sphere ") is a thin layer above
the solar surface extending upwards for perhaps 10,000
kilometres.   It is so called because at times of total solar
eclipse, just before the moon completely covers the sun, this
layer is momentarily left visible and shows as coloured red.

It is a region somewhat like the prominences, which extend upwards from it, in which electrons and ions are recombining. In consequence it emits a light of peculiar sort—almost entirely in certain selected wavelengths or colours characteristic of the different substances in the layer. Its constitution is still a matter for debate, and physical conditions there are obscure since, in some respects, it has characteristics appropriate to a temperature much higher than that of the sun's surface: it seems illogical that a layer further removed from

Figure 32

Sketch showing solar surface with " grassy " chromosphere and faint corona; the prominences are moving into the cool funnel of the sunspot.

the central heat of the sun than the actual surface should have a temperature higher than that surface. The chromosphere again is a feature of the sun which is only observable outside solar eclipses with special instruments.

The second feature referred to is the corona, again only observable in the ordinary way at times of total solar eclipse, but, like the chromosphere and prominences it can be observed by the special techniques developed by Lyot, some of which have already been described. It is one of the curiosities of astronomical history that the corona must have been seen by literally thousands of people before it was realised that it existed. It appears as a pearly white radiance surrounding the sun and extending outwards to about one solar diameter from the edge of the disc. It is so faint that it only becomes visible when the moon cuts off the direct sunlight. Its total light is about one millionth of that of the sun, or about half that of the full moon. Its shape varies with the sunspot cycle, being almost uniform round the sun at times of sunspot maximum, but showing streamers from the north and south poles of the sun at times of sunspot minimum. Of recent years these coronal structures have been closely studied and attempts have been made to correlate them with features, such as sunspots, appearing on the surface of the sun, but without a great deal of success. The explanation of the failure to realise that the corona existed (at later times it was thought that it was an appendage of the moon) is probably explained by the fact that it resembles nothing so much as the sort of halo of light often seen round the sun or moon when it is obscured by cloud, and it was probably assumed that it represented no more than background light coming from the sun. This explanation is not tenable, because the presence of such background light presupposes the existence of a cloud of scattering material. This is not the case, space round the sun being effectively empty. (Figure 32.)

To return to the corona itself: if the solar atmosphere is a puzzle, the solar corona is a mystery, and the pieces of information about its structure which have been obtained serve, in some ways, merely to add to the obscurity. The corona contains atoms, but they are highly ionised (that is, each atom has lost many electrons) and for many years they were unrecognised in their denuded form. Now that they have been recognised as ordinary atoms of calcium, iron, nitrogen and so forth, the puzzle is to explain how they got into this stripped state, since such a degree of stripping is characteristic of a very high temperature indeed—a million degrees centigrade or more. But the corona does indeed appear to be at such a temperature, for it is possible to measure the average velocity of agitation of the electrons forming a considerable proportion of the particles composing the corona, and the degree of agitation indicates a similar temperature. The physical basis of such a measurement is that the higher the temperature of matter, the greater the degree of agitation of the particles—atoms, electrons and so forth—which compose it.   Increased agitation is inseparable from increasing temperature, and, in fact, to say that the atomic or molecular particles are moving faster is no more than another way of saying that the body containing them has become hotter.   If the corona were not hot, and if ions and electrons were present together, the electrical attraction between them would lead to the immediate capture of the electrons by the ions.   The high velocities of random agitation make it much harder for the ions to capture the electrons, and enable the material to remain in a dissociated condition.

The corona is continually radiating energy out into space, and its temperature and degree of excitation must, in some way as yet unknown, be continually maintained from below by the much cooler sun.   Just how paradoxical this situation is can be realised from a perfectly fair analogy: it is almost as if we put a kettle of water on a block of ice and found it merrily boiling away.

After this most cursory survey we must leave the sun: we have dealt with it only in so far as solar knowledge is related to more or less familiar terrestrial things, such as radio, and in so far as we shall need solar properties in the explanation of what is to follow.   It is now time to return to our brief and to discuss once more the kind of astronomy which one can teach oneself.

CHAPTER V

COMMON PHENOMENA

There is a considerable number of astronomical phenomena which can be observed without the use of any instrument except the eye. These fall into two convenient classes, common and rare.

Under the heading of common phenomena we shall include the zodiacal light, meteors, occultations and variable stars. Under the heading of rare phenomena we include eclipses, comets and novae.

The Zodiacal Light

The zodiacal light is a permanent feature of the sky but it requires somewhat special circumstances for its observation. In the last chapter we said that the space round the sun was empty, but, like many statements in astronomy this is not precisely true. In the plane of the ecliptic, that is the plane of the orbits of the earth, and approximately of most of the bodies in the solar system, there is a good deal of dust, with possibly a proportion of particles of more considerable size, in the space between the planets. This dust scatters sunlight and therefore can be seen as a very faint luminous cloud. It extends east and west of the sun, along the ecliptic, and because it is faint, the sun itself must be hidden if the zodiacal light is to be seen: the phenomenon is best observed in places where the sun is as far below the horizon as possible when a point a few degrees east or west of the sun is just on the horizon. This means that the best place for the observation will be one where the sun sets almost vertically, that is, in or near the tropics. However, even in temperate latitudes the light can be glimpsed. One should choose a dark clear night and try to see it from a place remote from all artificial light where the horizon is free of obstructions. It may then be possible to see one end of a faint oval patch of light extending from the position of the sun along the zodiac or ecliptic. It is probably easier to see before dawn than in the evening, because the whole sky is then darker, the cold of night will have settled weather conditions, and the eye will be more rested and more capable of seeing faint illuminations. It may be of help to use a trick familiar to all astronomers, that of averted vision. This depends on the fact that the human eye contains two types of sensitive elements. Those near the centre of the retina are sensitive to bright lights and can register colour. In daylight when one looks directly at an object it is these elements which are operating. The sensitive elements at the edge of the retina are responsive to faint illumination but do not respond to colour. The two types of element occur all over the retina but are more numerous at the centre and edge respectively. These edge elements come into operation when one sees something " from the tail of one's eye " and normally in daylight, if the object is interesting, one swings one's eye round until it is the centre of the field of vision; that is, until the centre of the retina is being used. But at night this process brings a less sensitive part of the retina into use, and it can readily be verified that a star which is just visible out of the corner of one's eye will disappear when one looks directly at it. Astronomers wishing to see a faint object therefore use averted vision, and deliberately look at faint objects out of the corner of the eye. This requires some practice, because our normal habits make it difficult for us to bring our attention to an object without at the same time bringing it to the centre of our field of vision. With practice one can, however, avert one's gaze without averting one's attention.

It is believed that the dust particles responsible for the production of the zodiacal light extend out from the sun to a distance greater than the distance of the earth, and there should, therefore, be a second very faint patch of brightness visible in the sky in a direction exactly opposite to the sun. This patch is called the Gegenschein (German: counter light) but it is extremely hard to observe. It should be looked for as a circular light patch several degrees in diameter on very dark nights round midnight, on the meridian where the ecliptic crosses it.

Meteors

Meteors, or shooting stars are extremely common phenomena and a systematic watch of the sky for even a period of a few minutes, may, at certain seasons of the year, show dozens of these objects.

The name " shooting star " is a misnomer sanctioned by usage, for although meteors do resemble stars which have fallen out of place, they are, in reality, of quite a different nature. Most meteors are of the order of the size of a grain of sand and owe their large temporary luminosity to their impact, at velocities which may be as much as 30 or 40 miles per second, on the atmosphere of the earth. Very detailed thought has been given to the exact way in which this "burning up" of a meteor takes place: we need not go into that at the moment, and the crude picture of small surface particles being torn off the surface of the meteor by the friction of the air as it passes through, will suffice for us. The detritus forms a tail of luminous gas which may remain for a few tenths of a second, or in exceptional cases very much longer, to make visible the path of the meteor. After this interval the gas trail cools down or is dispersed by air movements. All this normally takes place at great heights in the earth's atmosphere, roughly the same as those at which auroral phenomena occur (say 30–60 miles) and the light emitted by meteors can, when suitably analysed, be used to give information about physical conditions at these great heights. Meteors by the million bombard the atmosphere of the earth every day and night, and all but a minute proportion are consumed long before they reach the surface of the earth. This is a fortunate circumstance, but occasionally meteors of much greater size arrive and can reach the earth before being consumed. These are a great rarity judged by the total number of meteors which arrive in the atmosphere, but meteorites (as meteoritic stones are called) are fairly common objects in museums. They range in size from small pebbles, up to gigantic freaks weighing several tons. The largest known meteorite fell in 1910 in an uninhabited area of Siberia and laid flat the pine forests for miles around, each tree pointing neatly away from the point of impact, as the result of the air blast set up by its passage and concussion. This object must have been of a very great size, perhaps hundreds of tons in weight. Meteorites all contain a high proportion of iron and are generally divided into two classes, stone meteorites and iron meteorites according to the percentage of the metal which they contain. In Arizona there is a crater about three-quarters of a mile across which is presumed to be of meteoritic origin, and magnetic measures on the floor of the crater indicate the presence of abnormal buried iron masses. In other cases lumps of iron put to all sorts of uses have been recognised as meteorites: one, in Mexico, which was in the form of an iron loop was found being used as an anvil.

Most meteorites in museums are iron meteorites, even though these are believed to form only a small proportion of all meteorites which fall. The reason is that iron meteorites are much more easily recognised for what they are. They are black in colour with a smooth surface covered with shallow indentations, known, from their appearance, as thumb prints. They closely resemble the marks which one can make in wet clay by thumb pressure. The microscopic structure of the material of iron meteorites also shows a variety of characteristic patterns which enable the material to be identified as of meteoritic origin.

The origin of meteors is still somewhat mysterious and it is a matter of some doubt whether they all originate within the solar system, or whether some arrive in our system from outer space. It is possible to plot the shape of the track along which the meteor is moving if there is enough observational material available: if this is a parabola or a hyperbola, then the meteor must have come from outside the solar system. If it is an ellipse, then the meteor has been moving round and round the sun like a minute planet until it accidentally encountered the atmosphere of the earth. The difficulty is that these three curves closely resemble one another over the small arc which is observable, and it is never possible in any individual case presenting any doubt to assign the observed curve to one type or another. Opinion seems to favour the idea that all meteors originate within the solar system and that they represent, as it were, the remnants of the building processes of the planets, or the fragments of some disrupted planet.

In one case it is possible to derive certain definite information about the origin of meteors. A great many meteors seem to appear almost at random, coming into the atmosphere from all directions. Those which happen to be moving against the direction of the earth's rotational motion come in very fast: those overtaking the earth come in at a much smaller relative velocity. But in addition to these sporadic meteors, *meteor showers* are observed. The properties of these showers are consistent with the idea that there exist in space streams of meteors travelling along elliptical paths round the sun. When the earth happens to cross through such a stream large numbers of meteors will be seen, and all of them will appear to be radiating from a certain point in the sky. This is an effect of perspective, because all the meteors in a stream are moving parallel to each other in space, but as observed, they appear to be radiating from a given point, much as the light rays from the sun (which are all parallel) appear to be diverging from the sun's position when the rays are seen coming through holes in a layer of cloud. (Figure 33.) This point (called a *radiant*) is actually in the direction from the observer parallel to the velocity of the meteor relative to the earth.

The origin of these meteor streams is well understood. Certain comets which moved in orbits round the sun, and reappeared periodically after a fairly constant number of years, have, in the past, disappeared. This has been due to the progressive break-up of the comets into smaller and smaller fragments, which, in course of time, have become spread out into a stream, tailing each other round and round the sun, and marking the elliptical orbit which the comet once followed. In one case, that of Biela's comet, this breaking-up process, or rather the initial division of the comet into two

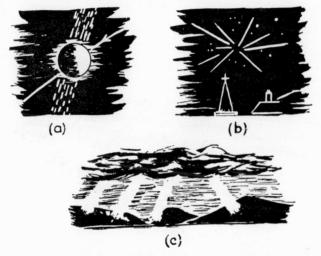

(a)   (b)

(c)

Figure 33

parts was actually observed. Later these large fragments must have disintegrated still further for eventually the comet was replaced by a recurrent meteor shower, appearing each year at about the same date when the earth passed through the stream of fragments. The existence of a number of these streams is well known and they have been given definite names. For example there are the Lyrids about April 21st, the Perseids about August 12th, the Orionids about October 19th, the Geminids about December 12th and several more. The Perseids and Geminids give the most abundant meteors. It is to be expected that the earth will pass through some streams during daylight so that they will be unobservable in the ordinary way, and this expectation has been confirmed by recent work using radar methods. The ionised tail left by each meteor is capable of reflecting a short-wave radio wave and of giving a pulse on a radar receiving set. By sending up beams of short-wave radio waves, workers at Manchester were able to detect a large number of pulses at certain times, and these have been identified as a meteor stream encountered by the earth during daylight.

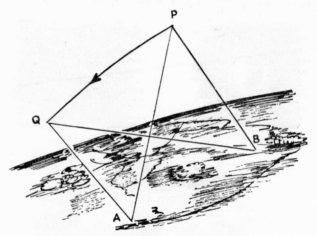

Figure 34

The observation of meteors is the field par excellence in which the amateur without any apparatus can render important services to astronomical research. A meteor passes above the earth at a relatively small height, say 50 or 60 miles, so that an observer at one place will see it in an entirely different direction from a second observer only a hundred miles away. These changes in the apparent position, length and direction of the path are the means whereby the true track in space may be determined. (See Figure 34.) A meteor observer must mentally note in the few tenths of a second during which the path is visible, the following data: the position of the beginning of the path, the position of the end of the path, and the time of flight. To do this he must be able to identify every visible star in the sky, for his method of marking the position of the beginning and end of the path is to note its position relative to the nearest visible stars. Then he can mark this position on a chart and read off its co-ordinates (right ascension and declination). The meteor observer sits or lies on his back in the most comfortable position possible, and notes down the data for each meteor which he sees, together with the time of the observation. The results of each night's work may be sent to one of the astronomical organisations (see list at end of volume) for analysis. The analysis goes something like this. Observer A at town X saw a meteor of great brilliance at 8.50 p.m. on January 25th. Observer B at town Y saw a faint meteor at 8.49 and a brilliant one at 8.51. Observer C at town Z saw a brilliant meteor at 8.48. The probability is that the watches of the observers were not quite accurate, and that the brilliant meteor seen by A, B, and C is the same object. All agree that it was travelling roughly from east to west. Assuming it is the same meteor we compare the data from the three sources. A calculation shows that A and B together make the starting height of the track 52·0 miles while B and C together make it 55·3 miles. In the end, the solution which fits the facts best may be that the starting height was 53·2 miles, the finishing height, 36·4 miles, the speed 27 miles per second and the direction 3 degrees north of west. So, three observers have contributed to exact knowledge of yet another meteor. It may sound a little pedestrian, but it is a piece of that priceless and irreplaceable commodity, reason-ably accurate information about nature. There is no other way to obtain it except by this rather laborious procedure, and if you should feel inclined to try your hand at this kind of work there are plenty of organisations which will welcome your aid. The greatest meteor observer who has ever lived was an amateur, the late W. F. Denning, who lived in Bristol.

Occultations

The observation of occultations is another field of re-search in which the amateur can do good work, but a small telescope and a reliable clock or stop watch are almost essential. An occultation occurs when the moon in its monthly motion round the sky passes in front of a star. The stars which are occulted lie near the ecliptic, but the motion of the moon is so complex and the rate of travel round the ecliptic of the node of its orbit (the point where the orbit cuts the ecliptic) so rapid, that a rather wide range of stars is occulted at one time or another. The chief piece of information to be obtained from an occultation is an exact location of the moon at that particular instant, and the reason why this information is important is that it enables astronomical computers to determine small error terms in the motion of the moon which, over the course of years may steadily build up into something quite appreciable. There is a certain source of error in the timing of occultations in that the surface of the moon is not smooth and, if a mountain should happen to be sticking up at just that part of the moon's limb which is covering the star, then there may be a difference of several seconds between the actual time of the occultation and that predicted for the smooth moon.

Again there is a considerable importance in the observation of occultations by observers scattered over the surface of the globe. The moon is relatively near to us, and a change of position from the northern hemisphere to the southern may correspond to a change in the apparent position of the moon against the background of the stars which is sufficient to produce an occultation as seen from the southern hemisphere when in the north the moon just grazes the star, or vice versa. Again, longitude is important: it takes rather more than an hour for the moon to cross completely in front of a star, and the whole process may have been completed as seen from one longitude before the moon has risen as seen from a place further west. Thus it is desirable to have observers equipped for observing occultations at places scattered all over the world, giving a coverage more complete than that which can be given by the professional observatories.

Predictions of the times of occultations as seen from various parts of the world are given in advance in the *Nautical Almanac* and tables are included so that an observer near one of the standard stations can compute very easily a corrected time for his own location. Two points dealing with additional refinements are worth mentioning. Work is now being done on the profile of the moon's limb at various conditions of libration, so as to permit corrections to be made for slight alterations in the time of an occultation produced by mountains or valleys at the moon's edge. Secondly, in the ordinary way, since the moon possesses no atmosphere to cause blurring or bending of the rays of light coming from the star, the extinction of a star or its reappearance should be quite instantaneous. There is a certain optical effect which should increase this time to about 20 thousandths of a second, but, in general, all stars should disappear and reappear in this time   However, it is believed that certain stars—bright stars of large diameter—should present a detectable disc, too small to be seen in any telescope but sufficiently large to make the time of extinction in these cases about one-tenth of a second. In particular during the past years there has been a series of occultations of Antares (α Scorpii), a star which is too far south to be occulted except when the moon is abnormally far south of the ecliptic. The period occupied by the extinction of light of Antares has now been timed and has been found to be about 0·10 seconds, corresponding to an angular diameter of this star of about 0·04 seconds of arc.

For the amateur with a small telescope of, say, three inches aperture, the procedure for timing an occultation is the following. First the expected time is found by looking up the tables and if necessary applying a correction. Then a stop watch is synchronised with true Greenwich or local standard time by pressing the button on the last pip of the six-pip time signal at the last available signal before the occultation. Then, looking through the telescope the observer presses the button again just at the moment of disappearance or reappearance of the star and so determines the time interval from the last time signal.

Variable Stars

Variable stars are stars whose brightness varies for one reason or another. Some variable stars vary almost or nearly regularly in a definite period of time. The shortest known period is rather more than one hour: periods may extend up to several hundred days or longer. It is possible to divide variable stars into two quite definite classes according to the cause of the variation of the light.

(i) *Eclipsing Variables*

The first class, the eclipsing variables, owe their variation to the fact that what is apparently a single star consists in fact of two stars, so close to one another that they are not separately distinguishable. The two stars move round one another, describing orbits round their common centre of gravity, much as the planets describe orbits round the sun, except of course, that in the case of the double star the masses of the two components are usually more nearly equal than are those of the sun and a planet. *Binary stars* of this type are by no means uncommon, indeed, something like a quarter of all the stars in the sky are believed to be double. What makes an eclipsing binary star out of a system of this kind is that the plane of the orbital motion passes through the earth. Thus, from the earth we see the orbit edge on, and the two stars appear to be shuttling back and forth along a line. When this condition is realised, then at each circuit each star passes in front of the other as seen from the earth and cuts off part of its light. It follows then that the curve or graph showing the variation of brightness of an eclipsing binary star should be fairly easily recognisable by certain features of its shape. We should expect to see a brightness (corresponding to the total light from both stars) which remains constant for most of the period of the cycle of variation. Then, quite suddenly, there should be a rapid fall in brightness to a minimum which corresponds to the light from only one of the stars. Then exactly half the period later (that is if the orbits are circular) the second star should eclipse the first. The situation is slightly complicated by the fact that the two stars will not have the same radius, nor the same brightness per unit area of surface. This means then, that if star A is bigger than star B, then when A is in front of B, only the light from A will be visible. When the positions are reversed and B is in front of A, part of the light of A only will be cut off, while that of B will all be visible. The depths of the two minima in the curve of the light variation will therefore not be equal. It may often happen, for example, that a pair will consist of two stars: a small but very hot one, giving out a great deal of light per unit area, and a large and much cooler one, giving out so little light per unit area that its total brightness is less than that of the small star. The interpretation of light curves of eclipsing binary stars so as to deduce the radii and surface brightnesses of the two component stars is a matter of no little difficulty, particularly when the picture is less simple than that which we have sketched and when certain additional complicating phenomena are present. However, it may be instructive to work out what the light curve ought to be in a simple case, given certain data about the two stars involved. Then it may become clearer how the process can be reversed and how important data about the components of an eclipsing binary star can be deduced from the observation of its light curve.

Let us suppose that we have an eclipsing binary star with a period of 10 days; that the two stars move in circular orbits; and that the data for the two stars are (Figure 35):—

Star A: Luminosity (i.e. total light emitted) = 4 units.
Radius = 1 unit.

Star B: Luminosity = 1 unit.
Radius = 5 units.

The total luminosity is 5 units, and, except during an eclipse, this will be the brightness of the system. Suppose we reckon time so that zero hour is at the moment when star B is just centrally covering star A. (In real cases the orbit may be slightly tilted so that the stars do not appear to pass centrally over each other.) Then, at this moment we can only see star B, since it is larger than star A, and all we shall see will be the 1 unit of light which B emits. Therefore at this moment the luminosity of the system as we see it will be down to 20 per cent of normal, i.e. there will have been a decrease in

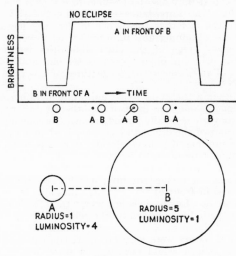

Figure 35

the brightness by about 1·75 magnitudes. Then, as time goes on, the stars will separate, but, after 5 days the position will be reversed, and star A will be centrally placed in front of star B. We shall see all the light from star A (4 units) and part of the light from star B. If star B is reckoned as a uniformly illuminated disc (this is not true in the case of the sun, nor for most stars): but, just supposing star B has a uniformly illuminated disc, then we shall see that portion of it which is not hidden by star A. Now star A has a fifth of the radius of star B, and hence one twenty-fifth of the area, so that when star A is in front of star B, we shall still get 24/25 of B's light. This fraction of the 1 unit of luminosity which B emits, gives us 0·96 of a unit of light, and this, together with the light from A gives a total of 4·96 units of light instead of the normal 5 units. Thus this minimum in the light curve represents a reduction of only about 1 per cent in the light or about 0·01 magnitudes, which is probably too small to be certainly observable. This is what happens in most cases of eclipsing binaries, but in cases where both minima are observable, it should be clear that estimates can be made of the individual contribution made by each component of the pair, and it may also be possible to estimate the ratio of their radii. The importance of eclipsing binaries may be judged by the fact that this is one of the very few methods available for obtaining any information whatever about the radii of stars. One of the chief difficulties is, of course, to know an eclipsing binary when we see one, and it is worth while to call attention to further features of this type of light variation.

If we assume the stars to be uniformly illuminated discs, then, as long as one is in front of the other, it does not matter whether they are centrally placed or not. As soon as one gets fairly in front of the other, the light received does not change. Thus the minima will have flat bottoms. Secondly, these minima should be of equal length unless the orbits are very elongated, for in the case of circular orbits with the stars moving round at a constant rate, it will take star A just as long to cross in front of star B as it takes star B to cross in front of star A. Lastly, it is important to note what proportion of the period is occupied by the eclipses, for this gives an indication of the sizes of the stars in relation to their distance apart. It the stars were very far apart compared with their radii, each eclipse would take only a small fraction of the period. If the stars were relatively very large—in an extreme case, if they were almost touching each other—then star A would start to eclipse star B almost as soon as star B had finished eclipsing star A. Obviously the details of the actual situation are susceptible of exact calculation from the observations, and in this way data of great importance have been obtained.

One naked-eye eclipsing variable star is the star Algol (β Persei, Map 2), an eclipsing binary star with a period of about 69 hours. At maximum it is of magnitude 2·3; at primary minimum about 3·5. The shallow secondary minimum gives a diminution of light of less than 0·1 magnitudes. The diminution of light at eclipse takes about 4½ hours from full brightness to minimum. The times of minima of Algol are given in *Whitaker's Almanack*.

(ii) *Intrinsic Variables*

The second class of variable stars is the class of intrinsic variables. Here we have to deal with a single star in each case, which, for reasons not at all well understood, varies its light output. The reason is to be sought in the structure of such a star, leading to a situation in which the atomic energy output at its centre, and the mutual gravitation of its parts are not in perfect balance. Hence, at some times there is too great an output of radiation, leading to the star's puffing itself up to a large radius. This is followed by a period of collapse, when the star contracts in on itself, pushing up the pressure and temperature at its centre, and stimulating once more an increased output of energy.

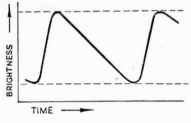

Figure 36

A considerable proportion of stars shows this kind of variability to a greater or lesser degree. Some are as regular as clockwork. Others exhibit irregularities of various sorts, ranging from mere variability of period or of brightness range to an almost complete lack of system. One type which is of great importance in astronomy are the Cepheid variables, a type of star of high intrinsic luminosity which varies with great regularity in a characteristic way, showing a relatively slow fall of brightness followed by a more rapid rise, and a period of fairly constant length. (Figure 36.) We may, if we like, think of the stars of this type as engines, pumping away quite regularly. The smaller, fainter stars pump very fast (in a matter of a few days) while the brighter ones are like heavy old-fashioned reciprocating engines with a far slower cycle of variation. This relation between intrinsic brightness and period of light variation is what makes the Cepheid variables important: they are extremely bright and so are visible even far out in the depths of space. A knowledge of its manner of variation enables us to recognise a Cepheid for what it is, however faint it may appear. Once it is recognised, the knowledge of the period enables us from previous researches to assign a definite intrinsic brightness to the star. Finally, the comparison of how bright it looks with how bright we know it to be intrinsically, enables us to estimate its distance from us.

The prototype of this class of star is Delta Cephei (Map 1), which varies between magnitudes 3·6 and 4·3 in a period of 5·27 days. Other naked-eye variables in the sky are Polaris, (the pole star) and Betelgeuse, in Orion; even the sun, with its 11 year cycle of variation may possibly be classed as a long period variable star. Another famous variable is o Ceti (Mira Ceti) (Map 2) which, at irregular intervals averaging 330 days, drops from a maximum brightness of about magnitude 3·5 down to about 9.

It is impossible to go into details of all the various types of intrinsic variable stars, except by writing several books on this subject alone. The short discussion of the Cepheids must suffice as an example. They are the type about which most is known and which are of the most use in other branches of astronomy, but the variables of other types present a tremendous field of research, which, so far, is dotted about with a host of somewhat unrelated discoveries, still awaiting integration into an ordered picture.

What concerns us now is that a few variable stars can be observed by the amateur even without a telescope if he knows where to look for them and how to set about it. However, he must remember that he can hardly do useful work on these objects and must treat the naked-eye variable stars merely as a hobby undertaken for his own amusement. Even a three inch telescope can, however, bring into view large numbers of fainter variable stars of which he can make useful observations which will be of value in research. The rôle of the individual amateur is again here a minor one: his work may eventually form part of a sizeable mass of observations whose analysis will lead to further understanding of the stars which he has observed. Addresses of organisations taking part in this work are given in the appendix.

Observing Variable Stars

The method of observing variable stars is simple enough,

but requires skill and practice to reach the greatest pitch of effectiveness. Select your variable star, or ask the organisation to suggest one, and either from a good star atlas or from a catalogue, or from your own observation, prepare a diagram showing what the field of stars round the variable looks like in your telescope. (Naturally one can do the same for the naked-eye variable stars.) You should choose stars covering a range of magnitude rather wider than the range of variation to be expected in the star. These selected stars are your comparison stars and should not be variables themselves, since the method is to compare the brightness of the variable star with that of the selected comparison stars. Supposing for example your star is expected to vary from magnitude 7·0

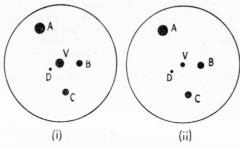

(i)   (ii)

Figure 37

to about magnitude 8·0 and that in the field of your telescope you can find stars of magnitudes 6·2, 7·6, 7·8, and 8·5. Call these A, B, C and D. (Figure 37.) You must try to imagine a certain number of steps of brightness between each of these— say 7 steps each of 0·2 magnitudes between A and B, 1 step between B and C, and 3½ steps between C and D. Then, making a note of the time at which you do so, you may observe that the variable star is intermediate in brightness between A and B and not much brighter than B. Comparing it with the brightness steps which you have imagined, you decide that it is two steps brighter than B. This means that you estimate the brightness as $7·6 - 2 \times 0·2 = 7·2$. A little later you find that the variable has dimmed and lies between C and D, and you then assign it a magnitude of 8·3. In practice, it is found that independent workers estimating in this way obtain results not very different from each other.

and that by this method one can really measure something fairly objectively and accurately.

Light in the Atmosphere

We conclude this chapter with some phenomena which can be observed without the aid of instruments. These are not strictly speaking purely astronomical phenomena but are concerned with the passage of light from the heavenly bodies through the earth's atmosphere. This is naturally a matter of the greatest interest to astronomers. Twinkling of the stars is caused by disturbances of rather small size in the atmosphere. One can, perhaps, think of them as columns or eddies of air, moving more or less independently of each other, each with a diameter of about a few feet. When starlight passes through one of these eddies it is diverted slightly from its original path, and hence there appears to be a small displacement of the star in the sky. This leads to the rapid and irregular changes of position and of brightness which we know as twinkling. Again, an air column of greater density than usual may cause a convergence or divergence of the rays coming to the eye. This refractive effect is more marked for the blue than for the red rays. In the case of very bright stars like Sirius, this separation of the light of different colours may be visible as flashes of different coloured lights in time with the twinkling. In a large telescope, of a sufficient diameter to collect light which has passed through a number of these columns of air, the effect seen is often not a shifting of a single image, but the superposition of a number of images shifting at random, so that, when the " seeing " is very bad, as it may be in a high wind, the star image is blown up like a balloon of fuzzy light.

The bending or *refraction* of light as it passes through the atmosphere is another well-established phenomenon. This takes place as the light passes from rarefied into denser layers of air, and is more marked when the length of the air path is greatest, i.e. when the star is fairly low down on the horizon. The effect is to bend the light rays slightly round the curve of the earth, so that the sun and stars seem at a slightly greater altitude than they would be in the absence of an atmosphere. (Figure 38 (a) (b)). This effect normally amounts to something like half a degree for a body on the horizon, so that the sun appears to set late and rise early, in consequence of this atmospheric effect, by about two minutes. Scattering and refraction of light in the atmosphere are the cause of twilight. Refraction is allowed for in tables of times of rising and setting of the sun and moon. Refraction also makes the horizon more distant than would be computed by a purely geometrical formula such as that given in Chapter 1. The refraction is not the same for light of all colours. The blue rays are bent more than the red. In addition the blue light is scattered preferentially by the atmosphere—as witness the blue colour of the sky. Normally these differential effects do not matter

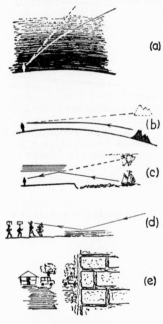

Figure 38

much, but when the sun is setting, interesting effects may be observed. To observe them a full unobstructed view of the western horizon is necessary, and care must be taken not to dazzle the eyes by looking directly at the setting sun. A cloudless sky at sea provides an excellent opportunity.

The differential refraction means that we see a red image of the sun, and a blue image of the sun at slightly different positions just at the moment of setting, with the red sun the lower. It may happen that the red sun sets completely while the blue remains still visible; or rather, the blue image would be visible were it not for the greater scattering of this colour. The red light being insufficiently bent by refraction, cannot continue to get round the bulge of the earth as long as the other colours. The blue light is scattered and merely goes to form part of the background. In favourable atmospheric circumstances an intermediate colour, green, may be left so that just at the moment of setting there may be a flash of green light on the horizon. It must be emphasised that this is not the general turquoise blue or green haze which is often seen above the horizon at sunset and which persists for many minutes. The green flash is a more dramatic phenomenon, sometimes taking the form of a transient plume of green fire which lasts only a few seconds.

This picture of refraction phenomena is a simplified one. Often there are complications due to the formation of layers of air of different densities—this can happen, for instance, near heated surfaces of stone or tarmacadam or over deserts. Then the refraction of horizontal rays may far exceed the normal value quoted above. It may be possible to see over hills which normally form the skyline, or, at sea, to have a range of visibility far in excess of that normally possible. If the air layers are well marked it may happen, according to a well-known optical law, that rays of light grazing either the top or bottom of a dense layer of air will be reflected there. Instead of seeing the desert sand we may see a reflection of the sky which looks like water (the mirage); ships on the horizon may appear upside down or part of a ship may be seen the right way up, the rest upside down. The same phenomena can often be observed on a very small scale. On the surface of the road we may see a shining reflection, or along the surface of a wall heated by the sun, we may, by looking along the wall with our eye only an inch or so from it, be able to see the layer of hot air in contact with the wall reflecting objects almost in line with it. This is no place to go into details. Suffice it to say that when anomalous conditions prevail on a large scale the phenomenon of the green flash may exhibit all sorts of variations and become much more marked and more easily observable. (Figure 38 (c) (d) (e)).

In the next chapter we turn from these fairly common phenomena to phenomena which are only observable on special occasions.

CHAPTER VI

RARER PHENOMENA

Eclipses

Eclipses are of two kinds: eclipses of the sun and eclipses of the moon. They take place when the sun, the earth and the moon are all in the same straight line, or very nearly so, so that the shadow of the moon falls on the earth, or the shadow of the earth on the moon. When the shadow of the moon falls on the earth there is a solar eclipse. Since the moon must then lie between the sun and the earth, the dark side of the moon must be turned to us: that is, eclipses of the sun always take place at new moon.

When an eclipse of the moon takes place the shadow of the earth falls on the moon: the moon must therefore be almost exactly opposite to the sun in the sky, and the fully illuminated face of the moon must be turned to the earth: that is, when an eclipse of the moon takes place the moon must be full.

If the orbit of the earth round the sun, and of the moon round the earth were in the same plane, that is, if the system could be exactly represented by a model in which the sun, moon and earth moved on the top of a flat table, then the earth's shadow would fall on the moon, and the moon's shadow on the earth once each in every revolution of the moon round the earth. There would thus be an eclipse of the moon every month at full moon, and an eclipse of the sun every month at new moon. The reason that this does not happen is that the orbit of the moon is tilted. In our model it should be represented as an oval drawn on a card which is tilted with respect to the table top at an angle of 5 degrees approximately, the top half of the oval being above the table, and the bottom half continuing down through the table to the lower side. The earth is, of course, at one focus of this oval curve. (Figure 39 (a)).

Thus it happens that at most full or new moons, the moon is above or below the table top, and no eclipse will occur. Eclipses will only occur if the moon happens to be crossing through the table top just at the moment when the three bodies are in line. These crossing points are called the nodes of the moon's orbit. The situation would still be relatively simple if these nodes were fixed points. If this were the case we could mark on a star chart the apparent track of the sun in the sky (the ecliptic), together with a second track representing the path of the moon against the sky background. These two fixed tracks would intersect in two fixed points (the nodes) which would be exactly opposite to each other in the sky. If the sun and moon, moving on these respective tracks, happened to be passing through the same node at the same

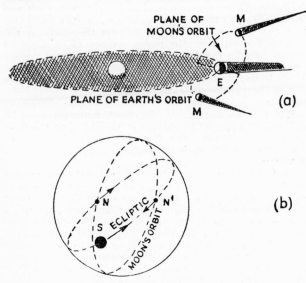

Figure 39

time, the moon would be in front of the sun, and there would be a solar eclipse. For a lunar eclipse to occur the sun and moon must be opposite to each other in the sky as seen from the earth. Hence, if the sun happened to be passing through one node, while the moon was passing through the other, there would be an eclipse of the moon.

Now the sun passes through various points on the ecliptic at almost exactly the same dates each year. If the nodes of the moon's orbit were fixed the sun would pass through them on definite dates in the year. Thus, in this simple imaginary system the prediction of eclipses would be easy since they would always occur close to two definite dates in the year, namely the dates of the passage of the sun through the fixed nodes, The fact that the nodes are moving makes the prediction of eclipses a much more complicated matter: there are regularities in the occurrence of eclipses but they are somewhat complicated.

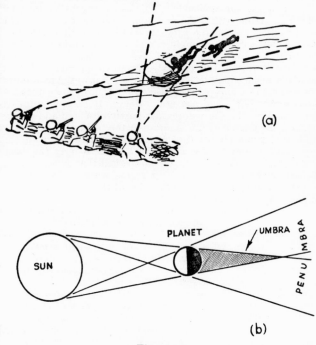

Figure 40

Before discussing them we must turn to a consideration of the various types of eclipse which can occur, and of the curcumstances attendant on them.

The Shadows of the Earth and Moon

The shadows of the earth and the moon are cast by the sun. The sun is an extended source of light which is much

larger than either the moon or earth.   It should be clear from
the figure (Figure 40) that these shadows will consist of two
parts.   First there will be a cone extending from the earth
or the moon away from the sun.   This cone is the *umbra*
of the shadow; it comes to an end in a point at a distance
determined by the sizes and separations of the bodies involved.
This is the zone of complete darkness, and from within it
no part of the sun is visible.   This can easily be verified by
taking any point within the umbra and joining it by a straight
line to any point on the sun.   It will be seen that this line,
along which a light ray might travel, is always interrupted by
the earth or the moon as the case may be.   Outside this
is a second region, the *penumbra* from within which part of
the sun's surface may be seen.   Take a point in the penumbra
and join it to various points on the sun : some of these lines are
interrupted, but some are not.   Hence, from any point in
the penumbra the sun will be seen partly obscured and partly
visible: the degree of illumination in the penumbra will vary
gradually, being almost that of the unobscured sun near the
outer edges, and almost complete blackness near the edge
of the umbra.   This corresponds to the fact that as one crosses
the penumbra one passes from a region where the sun's disc
is just nicked at the edge, to the edge of the umbra where the
sun is completely covered.

The nature of these shadows can best be illustrated by an
analogy.   The sun is an extended source of brightness, and
in the analogy we represent it by a trench manned by soldiers
armed with Sten guns each of whom pours out streams of
bullets over the whole arc of fire available.   The earth and
the moon are of smaller extent than the sun, and are to be
represented by a mound or a boulder out in front of the trench.
The problem of the observer plotting out the shadow cast by
the earth or the moon is the same as the problem of an enemy
soldier using the mound or boulder as cover.   The light rays
from the sun and the bullets from the Sten guns are both
travelling along straight lines.   The enemy is looking for
a shadow from the bullets, and he is safe when is is in the umbra
of the " bullet shadow " cast by the mound or boulder.   The
safe region is obviously limited for the following reason.   Since
the trench is longer than the width of the boulder the enemy
must not be too far behind his cover, because then he will be
vulnerable to cross fire from the ends of the trenches.   In
plan his safe zone will be a triangle extending back to a point
from the boulder.   This is the umbra of the bullet shadow
and the umbra of the light shadow is formed in the same way.

It is the dead ground behind the obstacle which cannot be reached by cross fire from the flanks of the defenders.

However, even if the enemy soldier is not in this safe zone, he can get partial cover to a greater or lesser extent. He may duck behind the obstruction so that it shields him from fire from one end of the trench, but he is then exposed to the fire from the other end. Clearly, if he goes too far off to one side he becomes completely exposed, but, as he moves in, provided he is not too far behind the obstruction, he will gain greater and greater protection. This is the penumbral zone of the bullet shadow. The degree of protection varies within it, and the degree of exposure to risk corresponds to the light intensity in the optical case, varying from full illumination at the edge of the penumbra, to full darkness at the edge of the umbra.

For any given relative positions of the sun, earth and moon, the dimensions of the shadows cast by the two latter bodies can be calculated. The nature of the eclipse phenomena observed in each case depends on the dimensions and location of these shadows.

Eclipses of the Sun

An eclipse of the sun takes place when the earth passes through the shadow cast by the moon. From what we have said about the inclination of the orbit of the moon it should be clear that the earth will not always pass centrally through the shadow. If, for example, the earth just enters the border of the penumbra, without intersecting the umbra, then from the shadowed portion of the surface of the earth the sun will be seen partially obscured. The degree of obscuration will vary according to the position of the observer on the surface of the earth. This is the most common type of solar eclipse, and is of relatively slight interest. The sun's disc is never completely obscured so that the faint corona and chromosphere cannot be observed. However, in recent years partial solar eclipses have acquired a new interest. It has been explained that the exact circumstances of an eclipse depend on the geographical location of the observer the relation may be inverted; observations of the circumstances of an eclipse may be used to provide exact data about the location of an observer. This may seem to be an aim which could be secured by other and simpler means. The continents have all been carefully surveyed and for each land-mass maps are available in which the locations are defined with very great accuracy. However, this is something which applies to each continent individually. Surveys of continents are very accurate, but the connections between continental surveys across sea barriers are far less satisfactory. For example, even in the case of so narrow a barrier as the English Channel, inconsistencies are present in the surveys on its two shores. Points in the south of England are included on French survey maps, and points on the north coast of France on the British Ordnance Survey, but the co-ordinates assigned on the two systems are not identical. Difficulties of this kind are aggravated in the case of more extended sea barriers. In order to reconcile such divergences it is necessary to make observations of some phenomenon visible over a wide area of the surface of the earth, which is capable of being defined in time with great accuracy. It has been suggested that partial eclipses of the sun fulfil these conditions, and observations along these lines have been made at recent eclipses. The particular feature which makes them valuable is that near the moment of maximum eclipse the direction of the line joining the points (the cusps) of the sickle-shaped area of unobscured sun is turning rather rapidly. The direction of this line can be accurately calculated and observations of its direction at various times throughout the eclipse can be made by taking photographs on ciné film. (Figure 42 (a).)

When the moon passes between the sun and the earth still closer to the line joining their centres, eclipses of a more spectacular character will occur. The circumstances will depend, as we have said, on the exact dimensions of the shadow cast by the moon. The sun has a diameter rather more than 100 times that of the earth: the moon's diameter is about one-quarter of the earth's. The dimensions are such that, on the average, the length of the umbra of the moon's shadow measures about 232,000 miles from the centre of the moon, but there is a considerable range on either side of this mean value. The average distance from the centre of the moon to the centre of the earth is 239,000 miles, that is, about 235,000 miles from the centre of the moon to the nearest point of the earth's surface. This distance is greater than the average length of the umbra of the moon's shadow, which means that, under average conditions, the shadow is too short to reach as far as the surface of the earth. Hence, even if the sun, moon and earth were exactly in line, no part of the earth's surface would intersect the umbra. In these circumstances an observer on the earth is, to revert to our analogy, too far behind the covering obstacle to obtain complete cover from the sun's fire although there is a partial cover. At the middle of the eclipse, the moon will have an angular diameter insufficient to cover the whole disc of the sun, and a ring of the solar surface will be visible all round. This ring gives this type of eclipse the name of an *annular eclipse*, a word which has nothing to do with the word meaning a year, but which is derived from the Latin word " annulus ", meaning, a little ring. (Figures 41 (b), 42 (b).)

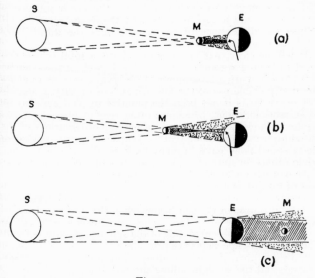

Figure 41

Total Solar Eclipses

Annular eclipses represent the average conditions of distance between the sun, moon and earth. Total eclipses of the sun, when the whole solar disc is blotted out, do occur, but require a conspiracy of favourable conditions of distance in addition to the almost exact collinearity of the three bodies. The orbit of the moon round the earth is perceptibly elliptical, so that the distance between the two bodies is variable. When the moon is nearer to the earth than the average at the moment of an eclipse, the umbra of the shadow may intersect the surface of the earth, producing a small region from which the sun may be seen totally obscured. (Figure 41 (a).) The maximum diameter of the shadow at the earth's surface is 167 miles, but since this distance is reckoned perpendicular to the axis of the shadow, these dimensions may be increased when the shadow falls at a relatively small angle i.e., when the sun's altitude is relatively small. The maximum possible duration of a total solar eclipse, that is, the time during which the sun is completely covered, or the time required for the umbra to sweep over the position of an observer as the earth and moon move past each other, and the earth rotates, is about seven and a half minutes. At the average total solar eclipse the dimensions of the shadow and the duration of totality are much less than these figures.

It will be clear that in order to observe a total eclipse one must take up a station on the predicted track of the shadow, and must prepare a programme of observations so as to make the best possible use of the few minutes of totality available. For many years it has been the practice to send expeditions to suitable locations, which are often remote places or uninhabited islands, for the purpose of making observations of the chromosphere, corona and other eclipse features. Many of these expeditions have brought back results of great value, while others, in spite of previous studies of the local climate, have met with overcast skies. North and south of the narrow belt of totality is a zone extending perhaps 2,000 miles in each direction, from within which the eclipse is seen as partial.

A total eclipse of the sun begins when the moon just begins to impinge on the sun's disc. This moment is known as first contact. The unobscured area of the sun is gradually reduced, and the landscape begins to grow dark and birds go to roost. The bright area of the sun grows more and more crescent-shaped, and the sunlight falling through small gaps in foliage imprints a legion of crescent-shaped light patches on the ground. (Figure 42 (d).) As the sun's visible area grows smaller it begins to approximate more and more to a star of unusual brightness. Ordinarily when even a bright star twinkles, the sole evidence available of the atmospheric disturbances through which its light is passing is afforded by the changes of colour and position. But the sun, now shrunk to a very small area, is still bright enough to cast a shadow, and evidence of the atmospheric disturbances which are always present, is available in the form of a vast number of shadow bands of irregular form and movement which can be seen playing on the ground and on the white walls of buildings. As totality approaches the umbra of the moon's shadow can be seen approaching across the landscape. Just before it reaches the observer the last rays of the sun can be seen shining down the lunar valleys, outlining the edge of the moon with a series of patches of brightness known as " Baily's beads ". (Figure 42 (c).) Then, as the sun disappears, totality begins.   This is " second contact ", the moment when the sun just disappears, and when the umbra of the shadow just reaches the observer. The pearly white corona, the chromosphere and prominences appear

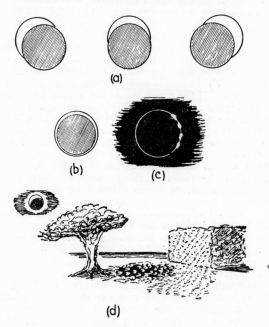

(a)

(b)   (c)

(d)

Figure 42

against the dark sky and some of the brighter stars are visible. No professional astronomer has time to glance at these phenomena, for now the work which has brought him thousands of miles has begun, and he is busy taking photographs according to the plan worked out months before and frequently rehearsed.

Just at third contact (i.e. the end of totality) the bright chromosphere is seen to flash out for a moment before the first edge of the sun appears, the corona disappears in the

E

renewed brightness, the sky brightens, and the eclipse is over, except for the rather dull business of the steady withdrawal of the moon from the sun's disc.

Lunar Eclipses

Total eclipses of the sun require rather special conditions for their production and must be observed from special very limited areas of the earth's surface. By contrast, eclipses of the moon are far less exacting. The length of the umbra of the shadow of the earth is about 859,000 miles, whereas the average distance of the moon is only 239,000 miles. Thus the length of the shadow is far more than adequate to reach to the moon, and the diameter of the shadow at the moon's distance is more than two and a half times the diameter of the moon. Thus there is no such thing as an annular eclipse of the moon. If the moon enters the umbra, there is a total or partial eclipse. Moreover, any eclipse which does occur is visible from any part of the earth's surface from which the moon is visible. The shadow is falling on the moon itself, and if an observer can see the moon, he can see that this is happening. (Figure 41 (c).)

When a total eclipse of the moon occurs, there is first a dimming of the moon as the penumbra is entered. Later the edge of the umbra is seen encroaching on the moon, until the moon is finally covered. Although to the naked eye the shadow seems fairly definite, to telescopic view the edge shows rather blurred. This is due to the fact that the sunlight at the edge of the shadow has passed through the atmosphere of the earth. Lunar eclipses are of only slight scientific interest now, but they do demonstrate, firstly that the earth is round, since it has a circular shadow, and secondly that the earth is roughly the size we know it to be. The word roughly must be included since in fact the shadow dimensions are slightly changed by the passage of the light through the atmosphere of the earth from what they would be in the absence of an atmosphere. It is, however, easy to see that the earth's shadow is much larger than the diameter of the moon.

When the moon is completely eclipsed there still remains a considerable light on the lunar surface, often of a coppery red colour. This light has passed through the atmosphere of the earth and its colour and intensity depend to an extent on the particular meteorological conditions which obtain on the rim of the earth visible from the moon. Thus the illumination is variable both as to colour and intensity from eclipse to eclipse, but, in general, the light has lost its blue components by scattering in the earth's atmosphere, leaving a rather eerie livid red colour.

When the moon passes centrally across the shadow the duration of the total phase is about an hour and three-quarters. The time from the first appearance of the umbra on the disc (first contact) to its final disappearance (fourth contact) is almost four hours.

The Recurrence of Eclipses

Since eclipses occur when the sun, moon and earth are in line or nearly so, this means that at such times the sun must be at or near a node of the moon's orbit, while the moon is near either the same node or the opposite one. The bodies, clearly, need not be exactly at the nodes, and hence the question arises, how near must they be for an eclipse to occur? The answers to this question in the various cases define what are called the *ecliptic limits* and they are different for lunar eclipses and solar eclipses. Without going into details it will be obvious that one of the factors which is involved is the vastly different rate of movement of the sun against the sky background as compared with the rate of movement of the moon against the sky background. The sun moves, roughly, one degree per day: the moon rather more than 12 degrees per day. The calculation is much the same as would apply in the case of two trains, one an express, one a very slow train, moving on two tracks which cross over at an angle of five degrees (the inclination of the moon's orbit to the ecliptic). What we want to know is the conditions under which a collision will take place, either engine to engine, or with one train side swiping the other. Given the relative speeds of the trains, their lengths and the angle of intersection of the tracks, what are the danger limits? The important point in the astronomical case is that the trains concerned are not to be represented by the apparent sizes of the discs of the sun and moon, but by the moon, and certain other circles. For example, take the case of a solar eclipse when the moon and sun are at the same node. What we wish to predict is whether there will be an eclipse at *some* point of the surface of the earth. This will happen if any point of the moon comes in front of any point of the earth, that is, if the moon cuts into the outside cone drawn in plotting the umbra of the earth's shadow. (Figure 41 (c).) At the distance of the moon this is a circle about ten thousand miles in diameter, or about five times the diameter of the moon. It is then readily found that an eclipse may take place if new moon occurs within 18½ degrees of the node, and certainly will occur if the distance is less than about 15 degrees.   (Figure 43 (a).)

In the case of lunar eclipses we have to consider the situation when the sun is at one node and the moon at the other as seen from the earth.   When this happens the shadow of the earth is directed towards the moon and has a diameter of about 5,700 miles at the moon's distance.   In finding the ecliptic limits for a lunar eclipse we have to think of the moon moving along its orbit, and a circle of the diameter stated moving

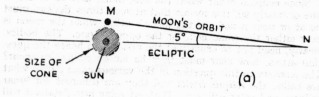

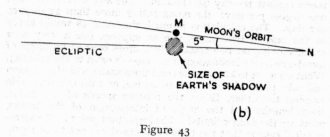

Figure 43

along the ecliptic at the solar rate.   They will overlap and an eclipse may occur if full moon takes place less than about 12 degrees from the node, and certainly will occur if the distance is less than 9½ degrees.   (Figure 43) (b).)

Two limits are given in each case because of variations in the inclination of the orbit of the moon, of the distance of the moon from the earth and other relevant quantities.

It is important to notice that, in the case of a solar eclipse there is a total angular region of 37 degrees (18·5 degrees on each side of the node) during which there is liable to be an eclipse.   At the rate of about a degree a day it takes the sun more than 37 days to pass through this zone. During this time the moon will have completed its circuit more than once, so that at least one new moon must take place within the ecliptic limits, and hence a solar eclipse must occur. The same is true at the other node. It then follows that in every year there must be at least two solar eclipses, total, annular or partial.

The sun actually meets two nodes of the moon's orbit in a period less than a year because these nodes are slowly moving so as to make a complete circuit of the ecliptic in 18·6 years. As we pointed out at the beginning of this chapter, if the nodes had been fixed, eclipses would always take place at two definite seasons of the year. Eclipses still take place at definite seasons corresponding to the passage of the sun through a node, but these are not fixed in the year. The nodes are moving to meet the sun in its motion round the ecliptic and the sun meets a node every 173 days instead of every 183. (Figure 39 (b).) The minimum number of eclipses which can occur in a year is two and both are solar, but, in exceptional conditions very many more than this can occur. For example, the sun may just have entered the ecliptic limits at the first new moon of the year, giving a solar eclipse. About a fortnight later the moon will have got round to its opposite node, and, if matters are properly arranged, full moon can occur within the ecliptic limits for a lunar eclipse. Then, a fortnight later the moon is back at the first node, where the sun will still be in the range of the ecliptic limit for a solar eclipse. Thus, under very favourable circumstances, there can be two solar eclipses within a month separated by a lunar eclipse. After the lapse of 173 days, there must be a solar eclipse when the sun reaches the second node. Then after a further period of 173 days (that is still within the calendar year) the sun returns to the first node and there is another eclipse. If all circumstances are favourable, as many as 7 eclipses can occur in a calendar year, 4 or 5 being solar, and 3 or 2 being lunar.

The Saros

To be exact, the sun passes through two nodes of the moon's orbit every 346·62 days, while the average length of the month from new moon to new moon is 29·5306 days. If we take 19 of the first of these periods and 223 of the second we arrive at a period of time which is almost exactly the same in the two cases—about 18 years and 10 or 11 days. The exact multiples are 6585·8 days and 6585·3 days. This period is called the *Saros*. If at the beginning of a Saros the moon, earth and sun were exactly in line, then, at the end of the period these bodies would be in the same relative positions, except for a relative movement of the sun corresponding to the difference of half a day between the two periods quoted above. In this time the sun moves only about half a degree (approximately its own diameter) against the star background. Thus, if the sun were within the ecliptic limits (as we have supposed) at the beginning of the Saros, it will again lie within them at the end. Thus, if there is an eclipse at a certain time, there is almost certain to be one closely reproducing the circumstances of the first, 18 years and 10 or 11 days later. Eclipses will go on reproducing themselves at intervals of a Saros until the half-day differences have accumulated to such a quantity that the ecliptic limits are exceeded. Corresponding solar eclipses in successive Saros periods do not occur at the same part of the earth. As the result of the difference of half a day, the consequent shift in the relative positions of the sun and moon, and the rotation of the earth on its axis, the tracks of successive eclipses are shifted westwards in longitude by about 120 degrees and also slightly north. Cycles of solar eclipses with as many as 70 members covering a period of nearly 1300 years, and of lunar eclipse with 50 members corresponding to a period of 900 years can be followed. The lunar series is shorter because the lunar ecliptic limits are narrower, and therefore sooner exceeded.

The Metonic Cycle

There is a further period which is of interest in connection with eclipses. This is a period of almost exactly 19 years formed of 235 months of 29.5306 days. We may contrast this with the properties of the Saros. The Saros leads to a prediction of the repetition of an eclipse of great certainty, but it does not occur at the same part of the earth, nor, since an incomplete number of years has elapsed, does it occur in the same part of the sky. Thus the Saros is of little use for the rule of thumb prediction of eclipses visible from a given place. Now, the period of 235 lunar months amounts to 6939·7 days, while 19 years amount to 6939·6 days. Thus if there is a new moon at the beginning of this period, there will also be a new moon at the end, and it will occur with the sun at almost exactly the same region of the sky. The shift of the sun's position in the difference between the two numbers quoted (one-tenth of a day) is only a fraction of the sun's diameter. However, owing to the inclination of the moon's orbit there will have been a relative shift of the two bodies north or south of the ecliptic. Now the sun passes through two nodes of the moon's orbit every 346·62 days (a period sometimes called an *eclipse year*) and twenty of these periods amount to 6932·4 days, differing from the previous numbers by about 7 days. In 7 days the sun moves about 7 degrees, while the total range of the ecliptic limits for solar eclipses is about 37 degrees. Thus it is to be expected that if the sun is just entering the ecliptic limits at the beginning of a nineteen year period, and there is an eclipse, there will be another one at the same time of the year in the same part of the sky nineteen years later, since the sun will have shifted only 7 degrees with respect to the moon's node and will still be within the limits. Allowing a shift of 7 degrees for each 19 year period it is clear that there will be eclipses on five or six occasions at 19 year intervals all at the same time of year and in the same part of the sky, after which the ecliptic limits have been exceeded and the series will be at an end. The 19 year period is called the *Metonic Cycle* and is useful for rule of thumb prediction of eclipses seen from a given place. Although the Chaldeans discovered the Saros the Metonic cycle must have been the basis of the methods by which the ancients predicted eclipses, but it is uncertain, since, without deeper knowledge one cannot tell when a new series of eclipses is liable to start, or, with certainty, when one is about to finish. This fact must have been the downfall of the ancient Chinese astronomers, picturesquely named Hi and Ho, who were executed for failure to predict an eclipse. The incident must have caused a pretty scandal, for the chronicle records that even the blind musician had beaten his drum, the mandarin had warned the people that the sun was being eaten up, and yet the professional astronomers were as indifferent as wooden images.

While it is thus possible to predict eclipses at a given place, it does not follow for example that all the solar eclipses will be total at the same position of observation. The totality belt of a total solar eclipse is narrow, and close calculation is required to fix it. Some of the eclipses of a series may be total, others partial or annular, and the totality belts of the total eclipses will be displaced over the earth's surface. A total eclipse of the sun at a given place is a great rarity, a fact which has had a considerable historical importance. On numerous occasions ancient chronicles have recorded a darkening of the sun, with the appearance of stars, coincident with some striking historical event, such as a battle, a plague and so forth. From the known geographical location it has often been possible to date the event, which on other evidence might be uncertain within decades, almost to a matter of minutes, even though it may have occurred in a civilisation with only the crudest knowledge of time-keeping and of calendars.

Comets

At the present time comets seem to be about as common as eclipses, for on the average something like half a dozen are observed each year. Most of these are so faint as only to be observable in telescopes, and as more telescopes and larger are brought into operation the number observed each year tends to rise.

In its full glory a comet consists of a nucleus, which probably contains all, or almost all, the solid material in the comet: round this nucleus is a faint haze of light, the coma, and stretching away from the head is the tail. The nucleus probably consists of a loose aggregation of solid particles or lumps of matter, possibly having a total mass equal to that of one of the smaller minor planets. No exact figures are available. The coma may have a diameter of a few hundreds or thousands of miles. The tail may be a few hundred thousand up to a few million miles in length. Comets move under the gravitational action of the sun: some of them have definite elliptical orbits, like those of the planets, but much more elongated, and therefore reappear at more or less regular intervals. Others make only a single appearance and are not seen again.

The tail, which is the chief glory of a comet, is absent until the nucleus approaches the sun. Then, by some mechanism not at all well understood, gas is gradually evolved from the material of the nucleus until it forms a cloud round about. Then the intense stream of radiation from the sun begins to exert a pressure on this gas. All light radiation exerts a minute pressure on matter: it is quite inappreciable in ordinary life, but is very important in many astronomical circumstances. When a cloud of gas in empty space, subject to no other forces, is acted on by radiation pressure, the results can be very spectacular. The gas moves out from the nucleus away from the sun and forms the tail of the comet. The comet's tail does not therefore trail behind the comet like the trail of a rocket: it is always directed away from the sun and thus sticks out from the comet in a slightly curved arc almost at right angles to the direction of its motion. (Figure 44.) As the comet nears the sun then, the tail grows steadily. Whereas at distances remote from the sun, the comet would be no more than a collection of lumps of rock owing any visibility to reflection of the sunlight from a very small surface, now, the development of the luminous coma and tail make it more readily visible. Almost certainly, it will be observed as it approaches the sun and grows a tail of steadily increasing length. Then, passing the sun it beings to recede, steadily losing its brightness until it can at last be seen no more.

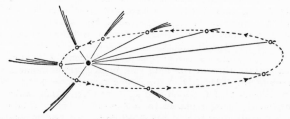

Figure 44

Diagram showing the orbit of a periodic comet round the sun. Features to note are the elongated shape of the orbit, the development of the tail near the sun, and the direction of the tail, pointing roughly away from the sun. The lengths of the tails are shown greatly exaggerated.

Certain comets are periodic, that is, they move on definite orbits round the sun and reappear a number of times. Easily the most famous of all is the comet called after Edmond Halley, the associate and friend of Newton, who applied the laws propounded by Newton to the prediction of the reappearance of the comet of 1682. This was a remarkable advance, not only because it represented a triumph for the scientific methods of Newton, but also because, for the first time, it brought comets, hitherto regarded as portents of dread events, utterly capricious in their behaviour, under the rule of natural law. The comet called after Halley has a period of about 77 years, the last appearance being in 1910. The period varies a little, since the orbit of the comet is apt to be slightly disturbed when it passes rather close to the planets, as may happen now and again. Previous to this, the comet had been observed at every reappearance since about 87 B.C. and its appearances have been recorded as far back as 240 B.C. The comet recorded on the Bayeux Tapestry, where the followers of Harold in 1066 are shown frightened by its appearance (the sempstresses did their best with the comet, but it does look like a combination of a sunflower and a flaming fork)—this comet is Halley's, occurring 844 years before its most recent passage near the sun.

Since the tails of comets are only temporary phenomena, then, even in a periodic comet, the appearance of the tail is liable to be very different from apparition to apparition. Some attempts have been made to determine, from studies of the shapes of the tails of comets, exactly what forces are at work, but these researches have not been very successful. Halley's comet, at its last appearance, had a very long tail, possibly 5 million miles in length or more, and the earth passed through it without ill effects.

Although Halley's comet has been extremely regular, it seems certain that in the long run, even periodic comets are impermanent. Already in discussing meteors we have referred to Biela's comet, which first split into two parts, and then disappeared, being replaced by a meteor stream. Doubtless only a minute quantity of gas is necessary for the production of the extremely tenuous tails of comets, and there is every evidence from the nature of the light emitted by the tails of comets that the gas pressure in them is extremely low—far lower than the pressure attained in even the best laboratory vacuum system on the earth. Nevertheless, the repeated production of a tail must involve some mass loss, and sooner or later it seems inevitable that comets must disintegrate and disappear.

In this respect then, there is only a difference of degree between the periodic comets and those which make only one appearance. A high proportion of comets are seen once as they pass the sun, and never again. It has even happened that comets have been seen very close to the sun at times of total solar eclipse and have never been seen either before or since. Comets doubtless end their lives either by splitting up or by having their orbits so disturbed that they fall into the sun or are captured by one of the planets. This last method of modification of comet orbits is well established. There is a whole group of periodic comets which at their greatest distance from the sun are at about the distance of Jupiter. The great mass of this planet has brought about disturbances in the orbits of passing comets until this state of affairs is reached.

If then, in either the long or short run, comets are gradually being done away with, where does the supply come from?

Recently Professor J. H. Oort of Leiden has advanced an interesting theory in which he supposes that comets are the residue of some hypothetical exploded planet. The fragments form a distant cloud round the sun far out in space beyond the most remote planet. Gravitational action of the nearest stars and of the members of the cloud with each other, leads every now and again to the changing of the orbit of one of these fragments into a highly elongated orbit bringing it close to the sun. This gives us a new comet, which may or may not be periodic. If it is periodic, then gradually the orbit will be changed by the action of the other planets, and the long process of degeneration will begin. The theory is rather speculative, but the ideas are of great interest to astronomers.

It will be clear that, in many ways there is not a great deal of difference between the nucleus of a comet and a rather small minor planet, and indeed it is possible that there is no hard and fast line to be drawn between these two classes of object, and, even with a large telescope, it is often somewhat difficult to distinguish between the two. It has been suggested that solar radiation pressure might produce a faint rudimentary tail on the dark side of the earth which might, for example, be responsible for the Gegenschein.

Novae

The literal interpretation of the name " Nova " is new (star). The phenomenon to which it applies is not the appearance of a new star, but the sudden increase in brightness of a previously faint one. Thus, if a nova is sufficiently bright, the casual observer may notice a star visible to the naked eye, which shortly before was absent. Many discoveries of novae have been made by people, amateurs of astronomy, whose work has compelled them to be out of doors at night. To make such a discovery requires not only extremely good luck, but also so exact a knowledge of the constellations that the appearance of a new member in one of them immediately catches the eye. The amateur is usually not in a position to exploit his discovery, but so great an astronomical importance attaches to the prompt recognition and report of a nova that we shall give a brief description of the nature of the phenomenon in the hope of increasing the number of potential discoverers.

The nova phenomenon seems quite capricious. Suddenly, for no known reason, a star will explode. Its brightness will rise in the course of a few days by a factor of more than ten thousand times, until a real brightness of something like a hundred thousand times that of the sun is attained. At stellar distances this will, usually, correspond to nothing more than the sudden elevation of a very faint star, perhaps to naked-eye visibility, perhaps to a somewhat fainter level. We now know that such a cataclysm is accompanied by the throwing off from the star of shells of very hot gas, at speeds of hundreds of miles per second. (Figure 45.)

After the explosion has reached its peak, the light of the star will die away gradually over a period of, perhaps, six months to a year. The dimming is not always very regular.

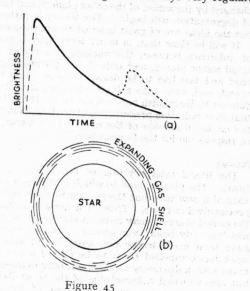

Figure 45

Sometimes there are minor fluctuations in brightness, and, in rare cases there may be a renewed outburst. The rate of appearance of novae, including instances too faint to be observed with the naked eye, averages something like one or two per year. Novae occur mainly in the Milky Way.

This catalogue of facts probably represents almost all the general information available about novae. The discovery that stars radiate by the production of atomic energy enables us to hazard fairly good guesses as to the general nature of the phenomenon. In general it is probable that a nova represents the results of a sudden failure of balance between the outward flowing stream of radiation which enables a star to keep itself distended against the mutual gravitation of its parts and this force of mutual gravitation. We have touched on such ideas in connection with variable stars, and have referred to the pressure exerted by radiation in connection with comets. Suddenly this balance becomes upset and there is a runaway production of radiation, analogous, perhaps, on a vastly greater scale, to the explosion of an atomic bomb. But having made this general observation we cannot claim to have advanced very far in our real understanding of the problem. Many questions remain : What kind of star is liable to this kind of explosion? We do not know, for most novae before explosion must be very ordinary-looking faint stars hardly likely to be singled out for special observation. Pre-explosion observations of novae are very rare. It is also rare for a nova to be observed during the short brightening stage, and a prompt report of the existence of a nova to the nearest observatory may enable observations of this stage to be made in time.

There is, in addition, a second class of nova, distinct from the first, known as supernovae. These phenomena are very rare. Superficially similar to the nova phenomenon, they are on a vastly greater scale with no examples intermediate between the novae and the supernovae. In a supernovae outburst the light emitted at maximum is equal to that from millions—on the average about 70 million—of suns. It is as if a star were suddenly created equal in brightness to all the stars of the Milky Way combined. Most of the instances observed have occurred in very remote nebulae, but two are known to have occurred in our own Milky Way, one in A.D. 1054 and one in A.D. 1572, both of which reached a brightness greater than Sirius.

CHAPTER VII

GET YOURSELF A TELESCOPE

Although it is possible to learn a good deal of astronomy without a telescope, the possession of even a very modest instrument is an immense advantage. The purpose of this chapter is to give a description of some elementary optical principles as they are applied to telescopes, and to offer some hints on the construction and use of the amateur astronomer's telescope.

The Refracting Telescope

Telescopes fall naturally into two classes, *refracting* or lens telescopes, and *reflecting*, or mirror telescopes, The refracting telescope is the type most familiar in ordinary use, for example, the telescopes used at sea, for shooting, and in the form of gunsights, are all refractors. Field glasses are merely a pair of refracting telescopes often of a rather special type of optical construction, mounted together so as to afford binocular vision.

The principal component of a refracting telescope is the large lens (the objective) mounted at one end of a tube, in the other end of which is a small lens or lens combination, forming an eyepiece through which the observer looks. Two optical dimensions of the objective are of special interest: they are its diameter, and its focal length. The diameter of the objective determines, among other things, the capacity of the telescope as a collector of light. In a well-constructed telescope, almost all the light caught by the objective reaches the eye. Now, at night, when the eye is accustomed to darkness, the maximum sensitivity of vision is attained. This is partly due to the opening of the iris, the ring structure carrying the colour of the eye which surrounds the dark aperture of the pupil. If the eye is regarded as an optical instrument it is the diameter of the pupil which limits the amount of light falling on the eye. The maximum diameter of the pupil is probably about a quarter of an inch, from which it follows that the total light from a sixth magnitude star which falls on a circle a quarter of an inch in diameter is just sufficient to excite the sensation of vision. When one uses a telescope with an aperture of three inches one collects all the light from a star which falls on a circle three inches in diameter, and, apart from losses by absorption in the lens and feeble reflection from the lens surfaces, all this light is funnelled into the eye. The three inch circle has a diameter twelve times that of the pupil, and an area which is 144 times as great. Hence, discounting the losses mentioned, the light available is 144 times as great, corresponding to a gain of rather more than 5 magnitudes. Hence the faintest star detectable is roughly of magnitude 11 instead of magnitude 6 with the unaided eye. A star of magnitude 11 is 100 times as faint as one of magnitude 6, so that if we have two stars of identical real construction which have these magnitudes, then the one must be at ten times the distance of the other. This follows from the fact that the intensity of the light received from a given source falls off as the square of the distance from that source. Thus, if we confine our attention to stars of a given type, the possession of a three inch telescope will enable us to detect such stars when they are at 10 times the limiting distance for detection by the naked eye. If we have two spheres of given radii, their volumes are proportional to the cubes of those radii. Thus the increase of range of ten times made available by the possession of such a telescope enables us to bring under observation a volume of space which is 1,000 times as great as that available to the naked eye. If stars were uniformly scattered through space this would mean that the number of stars observable would increase by 1,000 times, raising the observable number from about 5,000 for the whole sky observable with the naked eye, to about 5 million. The actual figures for the limiting magnitude for a 3 inch telescope, and the number of stars observable are respectively 11·2 and almost a million, the reduction in the latter figure being due to the fact that stars are not uniformly distributed in space. It will be clear that the slightest of telescopic aid does confer a tremendous advantage. It should be added that this argument based on the aperture of the telescope applies equally well to telescopes of any type provided that light losses within the instrument itself can be neglected.

The various types of lens may be distinguished by the convexity or concavity of their surfaces. The objective of a refracting telescope is equivalent to a lens of the biconvex type, that is a lens both of whose surfaces bulge outwards. A lens of this type, when held in front of a white card can form an image of a bright object in front of it. For example, a burning glass is of this type and it forms a small inverted image of the sun. If one experiments with a simple lens of this type it will be found that the lens must be held at a certain definite distance in front of the card if the image of a very remote object is to be sharp. For nearer objects the distance of the card from the lens must be increased. (Figure 46.)   This minimum distance is called the focal *length* of the lens and the ratio of the focal length to the diameter is called the *focal ratio*. If a lens of diameter 3 inches has a focal

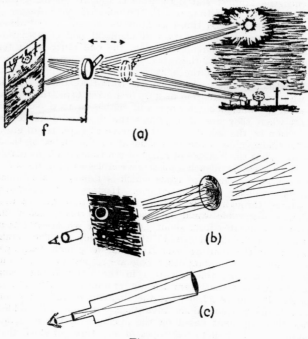

Figure 46

length of 30 inches, its focal ratio is 10, and we describe this by the symbol " f/10 ". We shall meet the same usage in connection with the concave mirrors used in reflecting telescopes.

The objective lens forms a focused image of the sky whose size is proportional to the focal length of the lens. For example, the telescope of 30 inches focus just considered will form an image of the moon about a quarter of an inch in diameter, and this figure will be doubled if the focal length is doubled. When a telescope is used for photography, a photographic plate is put in the plane of the focused image so that a picture of part of the sky is duly recorded. For observation by eye it is necessary to use an eyepiece, effectively a short focus lens used to magnify the image formed by the objective lens. In a refracting telescope this eyepiece is usually mounted in a tube of smaller diameter which can be moved in and out to secure a proper focus adjustment. It will be clear that the linear scale of the image formed by the objective lens is fixed, so that photographs with a given telescope are on a constant scale of so many minutes of arc to the millimetre. On the other hand the eyepiece, which is a magnifying glass in effect, may be changed so as to secure a greater or lesser degree of enlargement of this primary image. The greater the magnifying power of the eyepiece, the greater is the overall magnification of the telescope. The rule is that the magnification is equal to the focal length of the objective divided by the focal length of the eyepiece. Thus, an eyepiece with a focal length of half an inch will, when used in a telescope of focal length 30 inches, give a magnification of 60 times.

According to this rule it might be expected that unlimited magnification might be secured by reducing the focal length of the eyepiece. This is correct, but after a certain point the process ceases to yield improved results. The diameter of the objective determines the *resolving power* of the instrument, that is, its capacity for revealing detail. There is a rule that a telescope will just reveal two stars as separate when they are at an angular distance apart approximately equal to 5/D seconds of arc, where D is the diameter of the objective in inches. Thus, our three inch telescope may just resolve a double star into its components when these are about 1·7 seconds of arc apart. If they are closer than this, they will merely look like a single star, perhaps a trifle blurred, but not clearly separated. No amount of magnification will resolve them, and the use of greater magnifying powers will create a situation analogous to the case of a picture drawn on a rubber sheet, where stretching increases the scale but cannot reveal any new detail.

The capacity of the unaided eye for resolving detail is likewise limited. The eye can just distinguish as separate lines, two black marks on a white background whose angular separation is about one minute of arc, corresponding to a separation of about one inch at a distance of 100 yards. We shall clearly be approaching the maximum useful magnification of a telescope if we provide that the minimum resolvable separation set by the diameter of the objective, shall be magnified until it corresponds to the minimum distance resolvable by the eye. If we magnify more than this we merely stretch the image. If we magnify less than this we shall not bring up all the detail inherent in the image to a size at which it can be appreciated by the eye. To do this we must multiply 5/D seconds by such a factor that it equals 1 minute = 60″, that is we must apply a magnification of 12 for every inch diameter of the objective, or a power of 36 for our 3 inch telescope. This is actually only sufficient just to show all the detail possible, and in practice detail will be seen more easily if a higher power, up to, say, 50 magnifications per inch is employed. Still higher powers make the " rubber sheet " effect quite pronounced and begin to reveal in a rather marked manner the disturbances in image formation consequent on atmospheric disturbances. On the other hand if the magnification employed is too low, the bundle of rays emerging from the eyepiece has a rather large diameter, which may be so great that many of them fail to enter the pupil of the eye and go to waste. When this happens, only a portion of the objective is effective, and the best use of the telescope is not being secured.

So far we have merely considered the factors of objective diameter and focal length separately. Now we must briefly consider them in combination. The objective diameter fixes the amount of light collected. The focal length fixes the scale of the image. In the case of a single star the former is most important since it fixes the apparent brightness when seen in the telescope. No star is sufficiently near to be seen as anything but a point of light, and, provided the telescope is sufficiently well constructed so as to pack all the starlight into a single point, that is the end of the matter. This is never exactly true, since all telescopes suffer from defects of image formation inherent in the nature of light which cause even the most perfect telescope to image a star as a very small spurious disc surrounded by a number of very faint rings. These can sometimes be seen if a very high power is used on a bright star when the atmosphere is very steady. For the moment we can however neglect this aspect of the matter.

Thus the brightness of a star image depends solely on the diameter of the objective. The scale of the picture, and the apparent separations between various stars in a field of view depend on the focal length of the objective. However, matters are different when it comes to observing objects having each a perceptible extent, such as the moon, the planets or clouds of luminous gas in the Milky Way (nebulae). Then, as before, the total light collected depends on the diameter of the objective, but the size of the image formed by this light depends on the focal length. A short focus instrument will pack all the available light into a small image, which will, therefore, be relatively bright. A long focus instrument of the same aperture will spread the same light over a larger area and so produce a fainter image. This is of little importance for objects such as the moon and planets where the surface brightness is large, and where there is no risk that spreading out the image will make it so faint that it becomes difficult to see. For such work a large scale, leading to easy observation of detail is the first consideration; but when the object is faint, as in the case of comets and nebulae it is necessary to keep the surface brightness as high as possible in order to see them at all. The ideal to aim at is to keep the total light as great as possible (i.e. to make D large) and to pack it into as small an image as possible (that is, to keep the focal length small). These two requirements together mean that the focal ratio must be kept small. If the telescope is to be used for photography the smallness of the image is not of much consequence, since the photograph can afterwards be enlarged or examined under a microscope.

There is, of course, a limit to the smallness of the image. If it becomes too small so that it is indistinguishable from a star, it ceases to count as an image of an extended object, and the considerations applying to star images now apply to it. But as long as this limit is not reached, it is the focal ratio alone which counts in determining the brightness of the image of an extended object and not the diameter of the objective. Thus, strange as it may seem, the miniature camera working at $f/2\cdot5$ will be more efficient for the photography of, say a large comet, than the 200 inch telescope which works at $f/3\cdot3$. The difference is, of course, that the photograph taken by the latter will be much larger in scale, and show far more detail.

It may be asked why telescopes are not then made with much smaller focal ratios than they are. There are a variety of reasons, one of them being expense. The amateur photographer knows very well the difference in cost between a camera working down to, say, $f/6$ and one which works down to $f/3\cdot5$ and he also knows by experience how much slower the former is than the latter. The lenses in the short focus case have to be made with much greater curvatures and there are often more components in the lens. In the astronomical case there are often objections to the use of lenses with many components, quite apart from the fact that the manufacture of each component of a large lens is an expensive and tedious process.

Most astronomical objective lenses contain two components. These are lenses of different curvatures made of different types of glass. The reason for this type of construction is as follows: we have already met the phenomenon of refraction (the bending of light in a medium of varying density, or the sudden change of direction of light rays at the boundary between two transparent media). We have also noted that it is different for light of different colours. It would follow then that a simple lens, consisting of a single piece of glass would, to a slight but significant extent, separate out the light rays of different colours which constitute white light, and deviate them to varying degrees. Blue light is bent most, and red light least, so that the image in blue light would be formed nearer the lens than the one in red light. That is, the focal length for red light would be greater than the focal length for blue light. If a card were placed at the " red " focus a sharp red image would be seen, surrounded by a blue haze. The opposite would be true at the " blue " focus. It is, however, possible, by making the components of different materials, to produce a compound lens which focuses the blue and red rays at the same distance. There is a slight residual effect because the intermediate colours, yellow and green, are still slightly out of focus. In good cameras the lens often has three or more components which still further reduce this chromatic (colour) effect, but it is usually impracticable to follow the same plan for any but a very small telescope.

The Amateur's Refractor

Having dealt in some detail with the optical principles of telescopes, and particularly refractors, we may now apply these to the problem of obtaining a suitable instrument for amateur observation. It is possible for the amateur to construct a crude refracting telescope out of very simple materials. All that is necessary is a lens of fairly long focus, such as a cheap spectacle lens, a suitable cardboard tube, and a lens of fairly short focus, such as a cheap magnifying glass. The focal length of the spectacle lens is determined: this lens is stuck on the end of the cardboard tube as squarely as possible. The magnifying glass is stuck squarely on the end of another tube, preferably of a size to slide snugly in the first, and the large tube is cut off to such a length that a focus can be obtained and a certain range of movement on either side allowed. When properly focused the total length of the telescope will be equal to the sum of the focal lengths of the two lenses.

The construction is simple enough, and the telescope can be used for observing terrestrial objects. However, when such a crude instrument is turned on a star the need for greater refinement of construction will usually become apparent. The squaring-on of the lenses and the exact coincidence of their axes of symmetry is very important. If this is not secured, every star will possess a small tail like that of a comet. This defect is known as *coma*, and arises from defective centering and squaring-on of the lenses. To make a telescope suitable for astronomical observation requires a certain mechanical skill and access to workshop facilities. Local technical colleges may provide workshop courses which will afford the necessary instruction and facilities. Even then, the amateur is to be dissuaded from trying to manufacture his own lenses until he has gained some experience. If you are ambitious to construct your own refractor the organisations and books listed in the appendix will be of help.

However, a preferable course is for the amateur to buy a small telescope. There are, especially in large cities like London, a number of firms which sell secondhand astronomical telescopes, and even today something quite serviceable should be available at a price in the region of £30. The organisations listed will probably be willing to offer advice, but if you contemplate setting out on your own to buy a telescope there are a number of simple points which should be remembered and which will help you to lay out your money to the best advantage.

First examine the general mechanical condition of the telescope, and reject it if the tube shows signs of denting or bending or other damage The tube should be of stout brass and the objective should be mounted in a cell of its own which screws on with a very fine thread. The objective should have two components and you should examine it carefully for signs of scratching, bubbles or defects in the glass, or dirt. Do not take the components out of their cell: in a good objective the orientation and spacing of the components will have been carefully fixed by the makers, and you will destroy this adjustment if you remove them. The draw tube carrying the eyepiece should slide in and out perfectly smoothly without any play, and, if this is operated by a rack and pinion, the operation should be perfectly sweet. There should be two or three eye-pieces of different power going up to, say, 50 magnifications per inch diameter of the objective.

Test the telescope on a clear calm night and see that it produces a perfect image of a star. Test its resolving power by observing double stars like Castor, Mizar, γ Andromedae, the sword of Orion, clusters like the Pleiades and Praesepe. Use the formulae already given in this chapter to test whether it shows stars as faint as it should and resolves pairs of stars near the theoretical limit of resolving power. Look at the nebulosity in Orion and elsewhere and see how bright it looks, bearing in mind what has already been said about the connection between image brightness and the focal ratio of a telescope. Try in this way to form a critical judgment of performance and, if possible compare it with another telescope of fairly similar construction under similar conditions. You should now know enough to be able to allow for slight differences of diameter and focal ratio.

One of the most important features of a telescope is its mounting. All telescopes for professional use have what is called *equatorial mounting* the arrangement of which derives from the principles set out in Chapter I. For an observer in any given latitude the altitude of the pole is equal to the latitude. A telescope has to be mounted in such a way that it can move round two axes at right angles to each other. As an analogy, a gun is mounted so as to provide movement in elevation (i.e. movement about a horizontal axis) together with a traversing movement (movement round a vertical axis). The two axes need not be horizontal and vertical, and the principle of equatorial mounting is to make one axis parallel to that of the earth, i.e. to tip it up from the horizontal through an angle equal to the latitude. Swinging round this axis enables the telescope to be fixed in hour angle. Perpendicular to this axis is a cross axis, the declination axis, and movement round this corresponds to a change in declination. In a fixed observatory where the telescope is permanently mounted, this arrangement has great advantages. Scales fixed on the mounting enable the declination and hour angle of a position to be read off at once, and conversely a star can be picked up at once from the tabulated co-ordinates, right ascension and declination, even if it is far too faint to be seen with the naked eye. In addition, a driving clock is provided which steadily turns the polar axis at the diurnal rate, so that once an object is centred in the field it remains so. (Figure 50.)

However, the amateur will not usually require such refinements on his telescope. What he needs is an instrument which he can carry out into the garden when he wants to observe, and which can be stored in a clean dry cupboard in the intervals. The simplest type of mounting will suffice. This is the alt-azimuth type, a formidable name, which means no more than that the telescope can be turned up and down in altitude and swung left and right (in azimuth) to reach any part of the sky. It will, of course, be realised that objects must then be picked up with reference to nearby

Figure 47

stars, since the altitude and bearing of an object of given declination and right ascension are continually changing, and that the telescope will have to be moved by hand every few minutes to take account of these movements. This will, in practice, be found not to be a serious disadvantage.

An alt-azimuth mounting need consist of no more than a vertical pillar which can rotate, on the top of which is a plain bearing allowing movement in altitude. The telescope should be sufficiently well balanced, and the movements sufficiently tight to ensure that the tube stays in any position into which it may be put, and does not flop over. What is important about this type of mounting is that the

pillar shall be fixed on top of a tripod of sufficient size and
sturdiness. The tripod should be of heavy wooden construction with spiked legs which can be driven into soft ground.
It should be sufficiently tall to allow the observer to stand or
sit in comfort when using the instrument. (Figure 47.)
Any slight discomfort will be magnified to agonising proportions after quite a short period of observation. Some manu-
facturers produce what is called a " table stand " consisting
of no more than a pillar perhaps a foot in height, fitted with
three hinged feet. This type of stand is quite useless for
astronomical purposes and should be rejected out of hand.

The Reflecting Telescope

The principal component of a reflecting telescope is a concave
mirror whose upper surface is shaped to the form of a para-

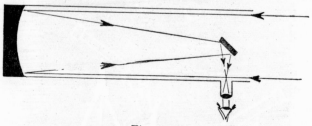

Figure 48

boloid (i.e. each section through the centre is a parabola),
this surface being covered with an almost microscopically thin
layer of silver or aluminium. This shape is adopted because
it has the property that all rays of light impinging on the
mirror parallel to its axis of symmetry are reflected so as to
pass through a definite point, the focus. (Figure 48.) It
must be noted that the light does not enter the glass of the
mirror, the sole function of which is to provide an accurately
formed surface to carry the metal reflecting layer. In a
refracting telescope the glass used for the lenses must, as far
as possible, be free from bubbles, inclusions and other blemishes. The glass for a reflector need not be of such high
quality, although of course, superlative quality is desirable.
The two components of the lens of a refractor have four surfaces which must be figured and polished. The reflecting
telescope has only one, although it must be shaped more
accurately than the surface of a lens. The reason for this

is that if the surface of a mirror is turned through a certain angle the reflected ray is turned through twice that angle. Hence, if there is a defect of shape in a mirror in the form of a small area tilted from the correct surface, this angular error will produce a doubled effect on rays reflected from that area. By contrast in the case of a lens, surface imperfections of this kind produce a reduced effect. In spite of this, the factors enumerated above—reduction in number of surfaces, less stringent requirements as to quality of glass—make reflecting telescopes of a given size much less expensive than refractors of equal size. A usual size for an amateur reflector is six inches aperture, whereas refractors of this size are unusual.

Most reflecting telescopes are of the Newtonian form. The main mirror is mounted at one end of a suitable tube, down which the light passes from the star. It is reflected at the mirror surface and returns in the form of a converging cone which would reach a focus on the axis of the tube just above its open end. It is impossible in small telescopes to put the eye in this position without blocking out the incoming light, so that a small flat mirror or prism is placed in the centre of the tube to reflect the light to the side where it enters an eyepiece or falls on a photographic plate. A small proportion of the incoming light is cut off by the shadow of the small secondary mirror and the thin vanes which support it, but the loss is inconsiderable. In the 200 inch telescope on Mount Palomar, California, and in the 120 inch telescope at Lick Observatory, also in California, it has been found possible to provide observing positions at the focus of the main mirror in the form of carriages suspended in the centre of the tube, into which the observer can climb.

The focal length of a mirror is the distance from the mirror surface to the focus. All the considerations of diameter, focal length, and focal ratio previously discussed, apply equally well to mirrors. The reflecting surface is apt to deteriorate with time. Silver surfaces rarely last more than six months especially if the air is contaminated with industrial smoke which usually contains sulphur. Aluminium surfaces last much longer, sometimes as long as ten years. Thus there is apt to be a falling off in the proportion of incident light reflected by the mirror, and for a reflector of a certain diameter and focal length there may be a loss of efficiency of perhaps as much as half a magnitude as compared with a refractor of the same dimensions. Two great advantages enjoyed by reflectors are, first, that all colours are reflected

F

alike from the mirror. There is no distortion of colour and the subtlest shades of colouring, for example on the surfaces of planets, are faithfully reproduced, Secondly, it is possible to make the focal ratio of a reflector relatively small without undue difficulty. Whereas refractors rarely have a focal ratio of less than 10, reflectors often have a focal ratio as small as 5. Further reduction of this figure is however a matter of difficulty since the curves which have to be figured on the mirror surfaces becomes inconveniently deep, and certain defects of image formation become more pronounced. We have remarked that a paraboloidal mirror brings all rays parallel to its axis to a perfect focus. That is to say, if the telescope is pointed exactly at a star a perfect image is formed. However, the telescope naturally has a certain field of view and stars at the edge of this field send in rays which arrive in parallel bundles at a certain angle to the axis. For such rays the focusing is not perfect, and at the edge of the field every star is imaged as a bright area from which a small tail or flare extends in a direction opposite to the centre of the field. This defect is called *coma* and cannot be suppressed in a reflecting telescope. (Figure 49 (d) (e).) It does, however, become more pronounced at a given distance from the centre of the field the smaller the focal ratio.

In a small reflecting telescope the mirror is supported in the simplest of cells. This need be no more than a dish-shaped piece of metal—an automobile brake drum is often used—in which the mirror can rest on a layer of thick felt or cork sheet. This serves the essential purpose of preventing direct contact between the glass and metal. In larger telescopes a special support system is provided. This usually takes the form of a number of sets of cork-faced pads, three to a set. The members of each set are mounted at the vertices of a triangular frame mounted on a rocking support. The use of three such sets gives support at nine points, while the rocking action automatically equalises the load on each support. In still larger telescopes, counterbalance weights are provided to afford sideways support when the tube and mirror are tilted. The purpose of these arrangements is to provide a sort of mechanical flotation system which will support the mirror without strain into whatever posture the tube may be turned. With a mirror weighing several tons there is a serious danger of distortion of the glass under its own weight and these support systems obviate the danger by providing an automatically adjusted support all over the back and sides. This is one of the reasons why all very large telescopes are reflectors:

a lens can only be supported at its rim, and beyond a certain size this limited support becomes inadequate to prevent sagging of the glass. Large mirrors on the other hand can be supported all over the back and sides.

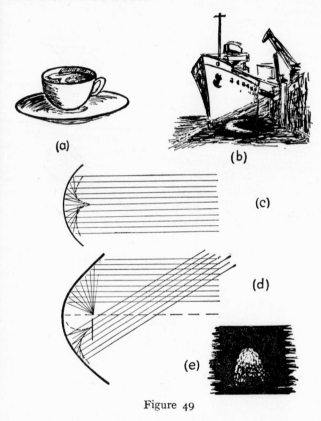

Figure 49

The mounting of a small reflector is usually rather squat and sturdy since the support of the tube is usually located quite near its lower end. This comes about since the support has to be near the centre of gravity of the tube, and the principal weight is provided by the mirror in its cell. For a six inch mirror working at f/8, the tube will be about four feet long, and the support may be about a foot from the lower end. A stand three feet high then brings the eyepiece about five feet from the ground when the telescope is pointed vertically.

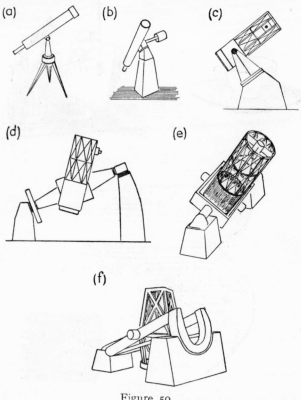

Figure 50

The observer can then stand comfortably alongside the telescope and easily observe the sky near the zenith where the best conditions of observing are usually to be found.

Large reflecting telescopes are always mounted equatorially. The polar axis may take the form of a large casting with the declination axis running through it at right angles. On one end will be the tube, on the other a large counterbalance weight. For very large telescopes the overhang of the weight of the tube to one side of the axis is something of a disadvantage and an alternative plan may be adopted. The polar axis may then take the form of a large frame with two parallel long sides joined by two shorter ones at right angles. The short sides carry the polar axis bearings and the tube is hung between the long sides where it can turn north and south. If the ends of the frame are straight the north-south movement is limited, so that it becomes impossible to point the telescope to the region of the pole of the heavens. In the 200 inch telescope this is avoided by replacing the northern short straight edge by a huge horseshoe bearing into the throat of which the telescope may be turned so as to reach the pole. (Figure 50 (f).)

In another system of mounting—the fork type—the polar axis corresponds to the handle of a two-pronged fork, while the tube is mounted between the prongs and turns in declination on bearings fixed to the ends of the prongs. (Figure 50 (c).)

The Amateur's Reflector

The manufacture of lenses is a somewhat difficult undertaking which should not be attempted without experience and special equipment. The manufacture of a concave mirror for an astronomical telescope is a process which requires patience but is not too difficult. Although small reflecting telescopes can be purchased through the channels already mentioned, it is quite common for amateurs to make their own mirrors, relying solely on instructions set out in books such as *Amateur Telescope Making*. Nothing special in the way of apparatus is required and an extraordinary variety of people, of both sexes, all ages and all kinds of occupations have found this kind of optical work interesting and pleasant. The principles are simple enough, but there are many possible refinements of detail. The purpose of the present brief sketch is merely to outline the methods, and, it is hoped, to stimulate interest to the point at which the reader may be inclined to make a closer study of the subject.

To begin, one requires two discs of glass, one of which will eventually become the mirror, the other being the tool. They should be of equal size and of the same material. Care should be taken that they are not too thin: a thickness not less than one-eighth of the diameter is desirable: for a six inch disc a thickness of one inch is ideal. Ordinary plate glass will serve, but pyrex, if obtainable, is to be preferred. Pyrex is much less expansible with change of temperature than ordinary glass, so that the exact shape of the mirror, on which its functioning depends, is much less vulnerable to temperature changes. The mirror is ground to shape by rubbing the tool over the surface of the mirror with a little wet abrasive placed between. There are only two pairs of surfaces which have the property that they may be moved one over the other while remaining in perfect contact in any position. One of these is a pair of planes: the other is a pair of spherical surfaces, one convex, the other concave, and of the same curvature. The purpose is to produce surfaces of the latter kind, the tool being worked to a convex shape, the mirror itself to a concave shape. The practical method of doing this is to lay the tool face up, preferably on the top end of a barrel, and to wedge it lightly but firmly in position with small wooden blocks. The barrel is suitable because it allows the operator to walk all round his work. It should be set up in some secluded spot, such as a shed or a garage, where it is possible to exclude dust and to hose down the work and its surroundings with water.

Some workers attach half a croquet ball to the back of the mirror by means of pitch. This serves as a handle on which a fairly considerable pressure can be exerted. Other workers simply use bare hands. To begin, a small quantity of coarse carborundum powder is put on the tool and wetted. The mirror is then laid face down, and a stroke is taken by pushing the mirror back and forth once straight across the centre of the tool. The next stroke is taken at a slight angle to the first, the worker having shifted slightly sideways. He repeats the operation throughout the work, shifting steadily round and round the mirror. As the process has been described, each stroke comes on the same diameter of the mirror, so, every few strokes the operator turns the mirror slightly to expose a new diameter to the abrasive action. The principle of the work is that on the overhang at the end of each stroke, the pressure on the edge of the tool is increased since the weight of the mirror is being carried on a relatively small area of the tool. Correspondingly the greatest pressure on the mirror at this point is falling on its centre. There is thus a tendency for the edge of the tool and the centre of the mirror to wear away, the mirror becoming concave and the tool convex. The process of walking round the barrel, and of slightly rotating the mirror, ensures that the wear is evenly distributed. It will be found that, quite rapidly, the mirror starts to become concave and the tool convex, the surfaces having the appearance of rather rough ground glass. After several charges of coarse carborundum have been used up, it may be judged that the work has gone far enough to change to a new grade. Not only must the surfaces be of the right form: they must also end up by being perfectly polished. This ideal is approached by stages. After the coarse carborundum, the work is thoroughly washed so as to remove all traces of the coarse abrasive and a finer grade is substituted. Work with this grade must be continued until all the pits made by the coarse carborundum are ground out, and the smaller pits of the new grade substituted. This shortens the focus of the mirror, but less rapidly, and converts the surface to the appearance of relatively fine ground glass. At any stage the focal length may be tested by wetting the mirror, setting it up on edge in a darkened room and testing it with a lamp. When the lamp is placed close to the centre of curvature but a little to one side, the image of the lamp will be seen coincident with the source. This coincidence can be tested by moving the head slightly to one side. If the source and the image in the mirror appear to move exactly together the lamp is at the centre of curvature. If they show relative movement, the one which moves with the eye is more distant. The basis of this test can be readily understood from the fact that at the centre of curvature all the rays from the lamp are falling perpendicularly on the mirror surface and are being sent back along their tracks. The radius of curvature, i.e. the distance or the lamp from the mirror is twice the focal length.

This process is continued with ever finer grades of carborundum until the mirror has a very finely ground surface, and a focal length slightly greater than that required. The surface has now to be polished. This is done with jeweller's rouge on a lap made of pitch. Some pitch with a slight admixture of beeswax is warmed until melted and then poured on to the tool. The mirror is coated with rouge and water and pressed on to form the lap. This is allowed to cool and the mirror removed. A series of v-shaped grooves is then cut in the lap with a razor so as to form a pattern of squares. The centre of the tool must be just in the corner of one of the squares, since otherwise the mirror will polish in rings. (Figure 51 (a), (b), (c).)

The mirror is then polished on the lap, using rouge and water as abrasive, until it has a perfectly smooth surface. At this stage great care must be taken not to scratch the mirror. The smallest particle of metal allowed to fall on the lap will scratch it beyond redemption. Dust must be excluded, but the room must not be dusted since this merely raises the dust and deposits it on the lap where it will do most damage.

The mirror is now an exact sphere and has a polished surface. The shape of this surface is sensitive. Touching it with the fingers will transmit sufficient heat to the surface to expand the areas of contact into perceptible hillocks a few millionths of an inch high. Polishing will heat the mirror by friction and it must be left to cool before testing its shape. What is required is a paraboloidal mirror, that is one which is slightly deeper at the centre than a sphere. The difference between the two shapes is very slight, and to obtain the right shape requires an accuracy of working many times greater than that of ordinary engineering practice. To produce this shape further polishing is carried out with parts of the outer edge of the lap cut away. It is not appropriate to go into details here, but it may be of interest to describe the rather simple methods of testing which will show when the surface has been shaped correctly to the paraboloidal form.

The Foucault Test

The test to be described is called after Foucault the famous nineteenth-century astronomer and physicist of that name. To carry it out, the mirror is set up on edge, and a very small light spot is placed at the centre of curvature by the method already described. The best light source to use is a lamp placed behind an opaque screen pierced by a pinhole. If the pinhole is slightly to one side of the centre of curvature it is possible to place the eye next to it in such a position that the whole surface of the mirror is seen illuminated. The smallest scratches or surface blemishes will immediately show up, especially if the light is fairly brilliant. Suppose the mirror is perfectly spherical, and suppose we have a knife blade, preferably a razor blade mounted on a screw support so that the edge of the blade can be brought in slowly across the line of sight. The tests can be done with a table knife held in the hand, but it is difficult to prevent the hand and eye wobbling about. A properly mounted knife edge and a support for the chin or a fixed hole for the eye to look through help a great deal. (Figure 51) (d.)

If the blade is brought in beyond the point of concurrence of the rays the cone of rays will be cut off on one side, and a shadow will be seen advancing across the mirror. If the blade is brought in on the nearer side of the point of concurrence of the rays, the shadow will be seen advancing across

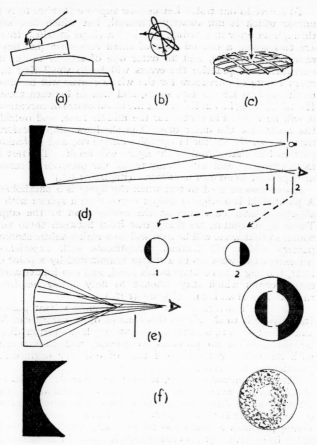

Figure 51

the mirror from the opposite side. If the blade is brought in exactly at the point of concurrence (the centre of curvature) the mirror will, if it is perfectly spherical, go dark all over instantaneously. A test of this type applied at the *focus*, using a bright star as the light source, is employed by astronomers to fix exactly the focal plane of a telescope.

F*

To revert to our test. Let us now suppose that we have a mirror which is not exactly spherical, but which has, say through an error in figuring, acquired a shape in which there are two zones, a central zone of small curvature (i.e. large radius of curvature) and an outer one of smaller radius of curvature. In practice the errors will be so small that the mean centre of curvature for the whole mirror will be close to the centres for the two zones, and will lie between them. If now we bring in a knife edge at the mean centre of curvature, it will be inside the centre for the middle zone, and outside the centre for the outer one. The shadows will, therefore, come across the mirror in opposite directions, and a shadow pattern like that shown in the figure will result. The test is most sensitive, and will readily reveal the presence of raised or depressed zones on the mirror. (Figure 51 (e).)

It can also be used to test when the figure is a paraboloid. A paraboloid is a slightly deeper curve than a sphere with a slightly greater curvature at the centre than at the edge. There is, of course, no sharp transition between these two zones, so that what will be seen will be a rather subtle shadow pattern which will become recognisable with experience. An expert can walk up to a mirror illuminated by a point of light, holding a table knife in his hand, and can immediately detect errors which may amount to only a few hundred-thousandths of an inch. (Figure 51 (f).)

This sketch covers only a few of the more salient points. Its purpose is to show that this class of optical work can be, and indeed often is, undertaken in ordinary home surroundings by people with no previous experience, and that results with an accuracy far beyond that of ordinary engineering practice can be secured.

When the mirror is made, it must be silvered: recipes for doing this are available (see appendix) or aluminised. The latter must be done professionally since it requires vacuum equipment. It is more expensive but more permanent. Silvering produces a much less robust coat, which is not only liable to tarnish, but also suffers much more if the surface is accidentally touched with the fingers. or if it is allowed to become wet with rain or dew.

The mounting of a reflector usually defeats the amateur unless he has workshop facilities, but, if these are available, there is a host of works describing methods of construction.

To summarise this discussion on the amateur's telescope we can say the following: If you want an instrument to amuse yourself sightseeing round the heavens and you are not mechanically minded, buy a refractor. If you are more ambitious and have more facilities, buy a telescope for the time being, and try making a mirror.

The Amateur Observer

When it comes to observing the amateur may do one of two things. Either he can amuse himself and his friends with tours round the heavens, wishing to go no further than to enjoy the strange and wonderful spectacles which are there to be seen. With the smallest of telescopes you may observe the moons of Jupiter and the markings on his surface. You may, by consulting the *Nautical Almanac*, be prepared in advance for transits of these satellites (passage in front of the disc) occultations (passage behind the disc) or eclipses (passages through Jupiter's shadow). You may observe the rings of Saturn, the polar caps on Mars, and the phases of the moon and Venus. You may, by plotting the tabulated position given in the *Nautical Almanac*, try to locate Uranus, and, when you have done so, verify the fact by observing the same position on several nights to see whether the suspected body has moved against the star background. You may observe the many double stars available, and delight in such spectacles as the Orion Nebula, the Pleiades, Praesepe, the Hercules cluster, Omega Centauri, and so forth. If you provide yourself with Norton's star atlas, you will soon learn your way round the sky as well as you know your own backyard. There is, in fact, a wealth of interest and amusement to be gained from such sightseeing. Most amateurs are content with this (as they have a right to be) and go no further. But, as we have explained, there are a number of types of genuine research which can be undertaken by the amateur even with a small telescope. Your local or national astronomical association will help you to take part in such co-operative programmes as the observation of double stars and of occultations, and, even if you have not got a telescope, you can take part in meteor observation. For work of this kind you need guidance more detailed than can be given here, and your association will be glad to provide you with it.

The Schmidt Camera

To conclude this chapter, we shall make a short reference to a type of instrument which represents something of a departure in telescope design, and which is receiving so much public notice that an account of it may help in the understanding of sometimes rather obscure newspaper reports.

The properties of a paraboloidal mirror are these: a bundle of rays parallel to the axis is reflected to a perfect focus, but a bundle arriving in any other direction comes to an imperfect focus, and shows the phenomenon of coma. (Figure 49.) The increasing importance of coma with angular distance from the axis, effectively limits the useful angular field of view. For example, in the case of a mirror working at f/5, coma is quite appreciable at an angular distance from the centre of the field which is a quarter of the diameter of the moon. All large reflecting telescopes of normal design have a very limited field due to this cause. Thus, although a large reflector giving a large-scale, detailed, image is ideal for the study of known objects of interest, the smallness of its field makes it unsuitable as an instrument for searching the sky for new discoveries. What is needed is an instrument which will very rapidly survey a large area of sky and which will pick up objects likely to be of interest for special study.

Now consider the properties of a spherical mirror. This has the properties that a parallel bundle of rays coming in in any direction is reflected, not to a perfect focus, but in such a way that all the reflected rays touch a certain curve. This curve is called a *caustic* and somewhat resembles two curved plumes springing from a single point rather in the manner of the two outer feathers of the fleur-de-lys. It is often formed naturally in familiar circumstances, the two best examples being provided by a cup of tea and a ship in dock. The milky tea shows very well the path of light rays through it, and the caustic can be seen well when a cup of milky weak tea is illuminated by a strong single source of light. The second case is best seen near the bow of a vessel moored alongside the quay in such a way that the sunlight shines on the bow plates near the water-line, the water there being in shadow. It is often possible to see a scimitar-shaped arc of light on and in the water in such circumstances, especially if the surface is perfectly still. (Figure 49 (a), (b), (c).)

Although a spherical mirror thus produces a defective focus at every part of its surface, all parts are identical in form. Thus, effectively the mirror has no special axis, and all over a very wide field the quality of image formation is equally good or equally bad. If a small diaphragm were placed at the centre of curvature, all parallel bundles of rays passing through it would fall perpendicularly on the mirror surface, and would be reflected to a rather poor quality focus lying on a curved surface having half the radius of that of the mirror. (Figure 52 (a).) If a film could be put there we should have a telescope producing rather bad images, but which would have an angular field many degrees in size, that is, possibly hundreds of times as large as that of a normal reflecting telescope. What is more, the focal ratio could be reduced, if not indefinitely, at least to values undreamed of for normal instruments. All that has to be done then is to find some way of keeping these desir-

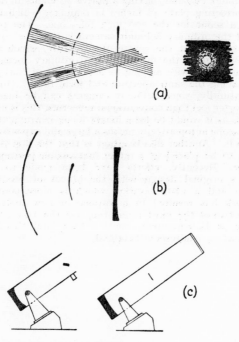

Figure 52

able properties while improving the quality of image formation. This difficult problem was solved first in the 'thirties by Bernhard Schmidt, an optician associated with the Hamburg Observatory. He designed a special kind of plate—the correcting plate—which is like a lens of rather slight curvature, formed into a double wave form (Figure 52 (b)) which will achieve this purpose. The manufacture of such plates is extremely tricky but the technique is rapidly advancing.

Schmidt cameras are now being used widely in connection with television, and in this case the plates are often cast from plastic, but for astronomical purposes something more exact is required.

In recent years a considerable number of Schmidt cameras has been constructed, one of the most notable being the huge one on Mount Palomar, having a mirror 72 inches in diameter and a correcting plate 48 inches in diameter which is being used as a scout for the 200 inch telescope. The principal defect of the ordinary Schmidt camera is that the correcting plate has to be at the centre of curvature, which is twice as far away from the mirror as the ordinary focus. That is, a Schmidt is twice as long as a telescope of conventional design having the same aperture and focal ratio. (Figure 52 (c).) Naturally, since Schmidt cameras can be made with much shorter foci than conventional reflectors, this is not quite as serious as it would be for a longer focus instrument, but a long telescope automatically means a large and expensive dome to house it. Another disadvantage is that the film is curved and has to be placed at a rather inaccessible position inside the tube. Recently, efforts have been made to modify Schmidt's original design with the object of producing a flat field and a plate position which is more convenient. This work has resulted in a variety of new instruments, of which two of the most interesting are the Baker-Schmidt telescope at Bloemfontein, South Africa, and the Linfoot-Schmidt at St. Andrews in Scotland.

APPENDICES

The Greek Alphabet

α alpha
β beta
γ gamma
δ delta
ε epsilon
ζ zeta
η eta
θ theta
ι iota
κ kappa
λ lambda
μ mu

ν nu
ξ xi
ο omicron
π pi
ρ rho
σ sigma
τ tau
υ upsilon
φ phi
χ chi
ψ psi
ω omega

The Constellations

Andromeda	Andromeda	*Cassiopeia*	Cassiopeia
		Centaurus	The
Antlia	The (Air) Pump		Centaur
Apus	The Bird of	*Cepheus*	The Sea
	Paradise		Monster
Aquarius	The Water	*Cetus*	The Whale
	Bearer		
Aquila	The Eagle	*Chamaeleon*	The
Ara	The Altar		Chameleon
Argo (Navis)	(The Ship) Argo	*Circinus*	The
Aries	The Ram		Compasses
Auriga	The Charioteer	*Columba*	The Dove
		Coma Berenices	Berenice's
Boötes	The Shepherd		Hair
		Corona	The
Caelum	The Chisel	*Borealis*	Northern
Camelopardus	The Giraffe		Crown
Cancer	The Crab	*Corona*	The
Canes Venatici	The Hunting	*Australis*	Southern
	Dogs		Crown
Canis Major	The Greater Dog	*Corvus*	The Crow
Canis Minor	The Lesser Dog	*Crater*	The Bowl
Capricornus	The Sea Goat	*Crux*	The (South-
Carina	The Keel		ern) Cross

165

The Constellations—(continued)

Cygnus	The Swan	*Pictor*	The Painter
Delphinus	The Dolphin	*Pisces*	The Fishes
Dorado	The Swordfish		
Draco	The Dragon	*Piscis Australis*	The Southern Fish
Eridanus	The Celestial River	*Puppis*	The Poop
Equuleus (Pictoris)	The (Artist's) Easel	*Pyxis*	The (Mariner's) Compass
Fornax	The Furnace	*Recticulum*	literally The Net
Gemini	The Twins		
Grus	The Crane		specifically the rhomboid reticle used by Lacaille in his meridian observations.
Hercules	Hercules		
Horologium	The Clock		
Hydra	The Hydra		
Hydrus	The Watersnake	*Sagitta*	The Arrow
Indus	The Indian	*Sagittarius*	The Archer
Lacerta	The Lizard	*Scorpio*	The Scorpion
Leo	The Lion	*Sculptor*	The Sculptor (his studio)
Leo Minor	The Lesser Lion	*Scutum (Sobieskii)*	The Shield (of Sobieski)
Lepus	The Hare		
Libra	The Balance	*Serpens*	The Serpent
Lupus	The Wolf	*Sextans*	The Sextant
Lynx	The Lynx		
Lyra	The Lyre	*Taurus*	The Bull
Mensa	The Table (Mountain)	*Telescopium*	The Telescope
Microscopium	The Microscope	*Triangulum*	The Triangle
Monoceros	The Unicorn	*Triangulum Australis*	The Southern Triangle
Musca	The Fly	*Tucana*	The Toucan
Norma	The Square	*Ursa Major*	The Greater Bear
Octans	The Octant	*Ursa Minor*	The Lesser Bear
Ophiuchus	The Snake-Strangler	*Vela*	The Sails
Orion	The Hunter		
Pavo	The Peacock	*Virgo*	The Virgin
Pegasus	The Winged Horse	*(Piscis) Volans*	The Flying Fish
Perseus	Perseus	*Vulpecula*	The Fox
Phoenix	The Phoenix		

Some Bright Stars

Star	Magnitude	R.A.	Declination
α Andromedae (Alpheratz)	2·2	0h 06m	+ 28° 49′
α Cassiopeiae (Schedar)	Var. 2·1–2·6	0h 38m	+ 56° 16′
β Ceti (Diphda)	2·2	0h 41m	− 18° 16′
γ Cassiopeiae (Tsih)	Var. 1·6–2·3	0h 54m	+ 60° 27′
α Eridani (Achernar)	0·6	1h 36m	− 57° 29′
α Ursae Minoris (Polaris)	2·1	1h 49m	+ 89° 02′
α Arietis (Hamal)	2·2	2h 04m	+ 23° 14′
β Persei (Algol)	Var. 2·2–3·5	3h 05m	+ 40° 46′
α Persei (Mirfak)	1·9	3h 21m	+ 49° 41′
α Tauri (Aldebaran)	1·1	4h 33m	+ 16° 25′
β Orionis (Rigel)	0·3	5h 12m	− 8° 15′
α Aurigae (Capella)	0·2	5h 13m	+ 45° 57′
γ Orionis (Bellatrix)	1·7	5h 22m	+ 6° 18′
β Tauri (Nath)	1·8	5h 23m	+ 28° 34′
ε Orionis (Alnilam)	1·8	5h 34m	− 1° 14′
ζ Orionis (Alnitak)	2·1	5h 38m	− 1° 58′
κ Orionis (Saiph)	2·2	5h 45m	− 9° 41′
α Orionis (Betelgeuse)	Var. 0·1–1·2	5h 52m	+ 7° 24′
β Aurigae (Menkalinan)	2·1	5h 56m	+ 44° 57′
β Canis Majoris (Mirzam)	2·0	6h 20m	− 17° 56′
α Carinae (Canopus)	−0·9	6h 23m	− 52° 40′
γ Geminorum (Alhena)	1·9	6h 35m	+ 16° 27′
α Canis Majoris (Sirius)	−1·6	6h 43m	− 16° 39′
ε Canis Majoris (Adhara)	1·6	6h 57m	− 28° 54′
δ Canis Majoris (Wezen)	2·0	7h 06m	− 26° 19′
α Geminorum (Castor)	1·6	7h 31m	+ 32° 0′
α Canis Minoris (Procyon)	0·5	7h 37m	+ 5° 21′
β Geminorum (Pollux)	1·2	7h 42m	+ 28° 09′
γ Velorum	1·9	8h 08m	− 47° 11′
ε Carinae	1·7	8h 21m	− 59° 21′
δ Velorum	2·0	8h 43m	− 54° 31′
λ Velorum	2·2	9h 06m	− 43° 14′
β Carinae	1·8	9h 13m	− 69° 31′
α Hydrae (Alphard)	2·2	9h 25m	− 8° 26′
α Leonis (Regulus)	1·3	10h 06m	+ 12° 13′
α Ursae Majoris (Dubhe)	2·0	11h 01m	+ 62° 01′
β Leonis (Denebola)	2·2	11h 47m	+ 14° 51′
α Crucis (Acrux)	1·1	12h 24m	− 62° 49′
γ Crucis	1·6	12h 28m	− 56° 50′
β Crucis	1·5	12h 45m	− 59° 25′
ε Ursae Majoris	1·7	12h 52m	+ 56° 14′
α Virginis (Spica)	1·2	13h 23m	− 10° 54′

Some Bright Stars—*(continued)*

Star	Magnitude	R.A.	Declination
η Ursae Majoris (Alkaid)	1·9	13h 46m	+ 49° 34'
β Centauri	0·9	14h 00m	− 60° 08'
α Boötis (Arcturus)	0·2	14h 13m	+ 19° 27'
α Centauri (Rigel Kent)	0·1	14h 36m	− 60° 38'
β Ursae Minoris (Kochab)	2·2	14h 51m	+ 74° 22'
α Scorpii (Antares)	1·2	16h 26m	− 26° 19'
α Triang. Aust.	1·9	16h 43m	− 68° 56'
λ Scorpii (Shaula)	1·7	17h 30m	− 37° 04'
α Ophiuchi (Ras Alhague)	2·1	17h 33m	+ 12° 36'
θ Scorpii	2·0	17h 34m	− 42° 58'
ε Sagittarii (Kaus Australis)	2·0	18h 21m	− 34° 25'
α Lyrae (Vega)	0·1	18h 35m	+ 38° 44'
σ Sagittarii	2·1	18h 52m	− 26° 22'
α Aquilae (Altair)	0·9	19h 48m	+ 8° 44'
α Pavonis	2·1	20h 22m	− 56° 54'
α Cygni (Deneb)	1·3	20h 40m	+ 45° 06'
α Gruis (Alnair)	2·2	22h 05m	− 47° 12'
β Gruis	2·2	22h 40m	− 47° 09'
α Piscis Aust. (Fomalhaut)	1·3	22h 55m	− 29° 53'

A Short Astronomical Dictionary.

Altazimuth.   Simple type of telescope mounting affording movement in altitude and azimuth (bearing).

Altitude.   Angular elevation above the horizon.

Angular separation.   Angle formed at the observer between directions to two stars or other objects.

Annular eclipse.   Solar eclipse such that a ring of the sun is left visible at maximum obscuration.

Aphelion.   Point on orbit of planet, comet, etc, most remote from sun.

Arctic, Antarctic, Circles.   Latitude limits of appearance of midnight sun.

Aries, First Point of.   The vernal equinox: Origin of co-ordinates of position on celestial sphere. The point, 0h R.A., Declination, zero. An intersection of the equator and the ecliptic.

Asteroid.   A minor planet.

Astronomical Unit.   Average distance of earth from sun. About 93,005,000 miles.

Atom.   Smallest unit of chemical substance.

Aurora (Borealis, Australis). High altitude display of light, electrical in origin, occurring near the north (south) polar regions.

Averted vision. Technique of observation with oblique gaze for detection of faint luminosity.

Baily's beads. Phenomenon occurring at the beginning or end of totality at a total solar eclipse due to sunlight shining down lunar valleys. The appearance is that of a string of beads of light fringing the moon.

Binary star. Pair of stars moving round each other under mutual gravitational attraction.

Caustic. Curve touched by system of rays in optical phenomenon, e.g. by reflected rays from a mirror.

Celestial sphere. Imaginary sphere of large radius used for marking directions to celestial objects. *Celestial poles:* points on celestial sphere in directions parallel to axis of the earth. *Celestial equator:* line marking projection on to celestial sphere of earth's equator. Line of zero declination.

Cepheid variable. A type of intrinsic variable star exhibiting characteristic and very regular cycle of variation.

Chromosphere. A solar surface layer of high temperature characterised by recombination of electrons and ions.

Circumpolar (star). A star which does not rise or set, and remains always above the horizon.

Coma (in a telescope). A defect of image formation leading to the appearance of spurious tails on all images. (Of a comet) luminous region surrounding the nucleus.

Comet. Body moving in orbit round the sun and characterised by the development of an extended luminous tail at closest approach to the sun.

Constellation. Arbitrary group of stars selected for convenience of designation of individual stars.

Corona. Solar appendage: a faint pearly white cloud surrounding the sun visible at times of total solar eclipse.

Cusp. The point or horn of a crescent.

Declination. Angular distance of a body north or south of celestial equator.

Earth Shine. Faint illumination of dark part of moon due to reflected light from the earth.

Eccentric. Literally, out of the centre. Used of an elliptical orbit in which the sun, occupying one focus, is at a distance from the centre.

Eclipse. Phenomenon which occurs when one celestial body passes into the shadow of another, as in the case of the earth and moon, or of Jupiter and one of its satellites, etc.

Eclipse Year. Interval required for sun to pass from one node of moon's orbit round to the same node again; about 346 days.

Eclipsing variable. Binary star in which the components as seen from the earth obscure each other during orbital motion.

Ecliptic. Apparent path of sun against star background.

Ecliptic Limits. Angular distances from node of moon's orbit within which sun must lie at full or new moon if an eclipse is to occur.

Electron. Ultimate particle of negative electricity.

Ellipse. Type of conic section. Form of all planetary and satellite orbits.

Equation of time. Difference between apparent time and mean solar time.

Equatorial. Type of telescope mounting with one axis parallel to that of the earth permitting direct setting in declination and hour angle.

Equinox, vernal, autumnal. See Aries, Libra, First Point of

Figuring. Process of working of optical component to exact shape.

First (second, third, fourth) contact. Used with reference to eclipses. First contact is the moment when the sun and moon first appear to touch. Second contact when totality begins or when, in an annular eclipse, the moon is first completely in front of the sun. Third and fourth contacts describe similar configurations (in opposite order) as the moon begins to pass off the sun.

Flare. Sporadic solar phenomenon involving sudden appearance of localised bright area on the sun, followed by terrestrial phenomena such as radio fade-outs, magnetic storms and aurorae.

Focal length, ratio. The focal length of a lens or mirror is the distance from the lens or mirror to the image of a remote object. The focal ratio is the ratio of this length to the diameter of the lens or mirror.

Focus. Point of concurrence of light rays. Point occupied by sun in relation to elliptical orbit of planet, comet, etc. The usage is actually identical in the two cases for light rays emerging from one focus of an elliptical mirror are concurrent at the other focus.

Foucault. 19th-century French astronomer and mathematician. The Foucault test for mirrors and lenses is named after him.

Gegenschein. Faint patch of light on ecliptic in direction opposite to sun.

Gibbous. Moon or planet having bright part greater than semicircle and less than circle.

Green Flash. Flash of green light sometimes seen momentarily at sunset when the horizon is clear. The phenomenon is due to refraction.

Greenwich Mean Time. Time system kept by clocks in Britian in winter. The Hour Angle of the Mean Sun at Greenwich plus 12 hours.

Greenwich Sidereal Time. Hour Angle at Greenwich of the First Point of Aries.

Gregorian Calendar. The modern calendar, with leap years every fourth year, except in century years whose first two figures are not divisible by four.

Hour Angle. Angle at north or south celestial pole between directions to a star, planet, etc., and meridian of observer. The value at any moment depends on the longitude of the observer; thence, *local hour angle*, *Greenwich hour angle*, etc.

Hyperbola. Conic section of one type (the others being the ellipse and the parabola). Orbit of object passing sun at very high speed, or of meteor passing earth at very high speed.

Ion. Atom deprived by any means of full complement of electrons. *Ionisation*, process of deprivation.

Julian Calendar. Calendar introduced in Rome by Julius Caesar, having leap years every fourth year. Superseded by Gregorian Calendar.

Knot   A speed of one nautical mile per hour.

Latitude. Angular distance, measured at centre of earth, of point north or south of terrestrial equator.

Libra, First Point of: The Autumnal Equinox, Point at Right Ascension 12h and zero declination. An intersection of the ecliptic and celestial equator.

Libration. Axial oscillation of moon relative to the earth.

Local Time. Time appropriate to longitude of place under consideration. Greenwich time adjusted for longitude difference.

Longitude. Angular distance measured along earth's equator of meridian through place east or west of Greenwich *Greenwich meridian*, line of zero longitude.

Magellanic Clouds. Two large aggregations of stars resembling detached portions of the Milky Way, in the southern sky.

Magnitude. Number describing the brightness of a star, planet, etc.

Maria. Smooth areas on surface of the Moon once erroneously thought to be seas.

Mean Sun. Fictitious body moving at uniform rate round celestial equator in one year, used as standard for defining Greenwich Mean Time.

Meridian. (1) Line on earth's surface running due north and south. (2) Line on celestial sphere running south from north celestial pole in northern hemisphere, or north from south celestial pole in southern hemisphere. *Meridian passage*, moment or act of celestial body crossing meridian as earth rotates.

Meteor. Small body rendered momentarily luminous by friction on entering atmosphere of earth. *Meteorite*, a meteor not consumed entirely which falls to earth. *Meteor shower*, group of meteors sharing common motion producing a display of numerous, almost simultaneous, meteors. *Meteor radiant*, position in sky from which all meteors of a shower appear to originate.

Metonic Cycle. Interval of 19 years at which a series of five or six eclipses may occur with the sun in the same part of the sky.

Midnight Sun. Phenomenon of continuation of daylight throughout the 24 hours occurring when the sun is a circumpolar star.

Milky Way. Irregular band of light spanning the sky produced by innumerable stars which constitute the star system of which the sun is a member.

Minor Planet. One of some 1,500–2,000 small bodies moving in orbits round the sun, mainly intermediate between those of Mars and Jupiter.

Minute of Arc. Angle equal to one-sixtieth of a degree. *Second of Arc:* one-sixtieth of this.

Moon. A satellite, especially the earth's satellite, describing an orbit round a planet under its gravitational attraction.

Nautical Mile. A distance of 6,080 feet corresponding to an angle of one minute of arc at the centre of the earth.

Nebula. Literally, " cloud ". Originally applied indiscriminately to any object seen as an extended patch of faint light. Now known to have been applied to many objects of different real constitution which are now distinguished as *extragalactic nebulae:* objects similar to the whole Milky Way: *gaseous or diffuse nebulae:* relatively near and small structures representing detail in the Milky Way. Also certain other types.

Node. Intersection of apparent paths of two objects, such as Moon's orbit with ecliptic, satellite orbit with planetary orbit, etc.

Noon, Apparent. Moment of meridian passage of sun.

Nova. Phenomenon of sudden explosive brightening of star.

Objective. Principal component of a telescope, i.e. main mirror or large lens.

Occultation. Passage of Moon or planet in front of star, or of planet in front of satellite.

Orbit. Path followed by planet or comet round sun, or satellite round planet.

Parabola. A type of conic section. Possible form of orbit of fast moving body under special circumstances. *Paraboloid:* surface of symmetrical form such that every section through the centre is a parabola. This is the form required for the surface of a mirror for use in a reflecting telescope. Such a surface is frequently described by mirror makers as " parabolic ".

Parallel of Latitude. Line of constant latitude; a circle on the earth's surface centred at the north or south pole.

Penumbra. (1) Region of partial illumination in shadow (of planet or satellite). (2) Zone of intermediate brightness at edge of sunspot.

Perihelion. Point of closest approach of a planet to the sun.

Period. Interval after which a phenomenon recurs, as period of a variable star, of orbital motion, of axial rotation.

Phase. Used to describe degree of illumination of moon, planet or satellite. Also in sense of fraction of period of a recurrent phenomenon which has elapsed, as for example, in connection with variable star. Also with reference to eclipses, to describe stage of phenomenon, as total phase, partial phase.

Planet.   Cool solid body moving in orbit round sun (or possibly star) owing light and heat to sun (or star). *Inferior, superior, planets:* those with orbits interior, exterior, to that of the earth.

Planetarium.   Device for demonstrating apparent motions of stars, planets, etc., usually by projection of images on interior of darkened dome.

Position Circle.   Circle centred at substellar point of star, planet, etc., having radius corresponding to measured zenith distance of object.

Prominences.   Large flame-like appendages of sun not normally visible without special instruments.

Reflector.   Type of telescope using a mirror for image formation. *Newtonian Reflector:* original form devised by Isaac Newton still widely employed.

Refraction.   Bending of light rays on passage from one medium to another, or into layer of same medium having different density, e.g. water to air, glass to air, hot air to cold, or vice versa.

Refractor.   Type of telescope employing lenses for image formation.

Relativity.   Theory of type proposed by Einstein which seeks to unify all physical laws in form independent of particular conditions of observation.

Resolving power.   Capacity of lens, mirror, eye for revealing detail.

Retrograde Motion.   Motion of planet, satellite, contrary to general direction of motion of bodies in solar system, or apparently so.

Right Ascension.   Number used in defining positions on celestial sphere analogous to longitude in terrestrial usage.

Saros.   Period of 18 years and 10 or 11 days after which eclipses recur, but not in the same part of the sky, or visible from the same region of the earth.

Satellite.   A moon.

Schmidt Camera: Type of telescope invented by Bernhard Schmidt having very wide field of view and producing very bright images of extended objects.

Sidereal day.   True period of axial rotation of earth relative to stars.   Equal to 23 hours 56 minutes of ordinary time. Divided into 24 sidereal hours, and these into minutes and seconds of sidereal time.

Sidereal Hour Angle. Alternative method of specifying Right Ascension adopted for purposes of aircraft navigation.

Sidereal Time. Time based on rotation of earth relative to stars. *Local Sidereal Time:* Local Hour Angle of First Point of Aries. *Greenwich Sidereal Time:* Greenwich Hour Angle of this point.

Solar Time. Time system based on hour angle of sun. *Mean solar time* based on hour angle of mean sun.

Solstice. Moments when sun is at greatest northerly or southerly declination. Positions where this occurs. As summer solstice, winter solstice.

Substellar, Subsolar, Point. Point at which star, sun, is in zenith.

Sunspot. Solar phenomenon analogous to meteorological disturbance in earth's atmosphere. *Sunspot cycle:* period of about 11 years during which frequency of appearance of sunspot varies.

Synodic period. Period after which the same relative positions of two planets, or of solar surface and planet, or planetary surface and satellite, are reproduced.

Transit. Passage of an inferior planet in front of sun's disc. A very rare phenomenon.

Tropical year. Average interval between successive passages of sun through vernal equinox.

Umbra. Zone of complete darkness in shadow of planet, satellite. Darkest part of sunspot.

Variable Star. Star whose brightness varies due to internal change or eclipse by a second star.

Zenith. Point on celestial sphere directly overhead. *Zenith distance:* angular distance from this point.

Zodiacal light. Faint illumination due to scattering of sunlight by dust, extending east and west of the sun along the zodiac (ecliptic).

Zone time. Standard time system adopted for convenience throughout a given geographical zone.

A BOOK LIST

GENERAL ASTRONOMY:
Astronomy, Russell, Dugan and Stewart, Ginn & Co.
General Astronomy, Sir H. Spencer Jones, Arnold.
The Heavens Above, J. B. Sidgwick, Oxford University Press.
Astronomy for Everyman, Martin Davidson, Dent.
Introducing Astronomy, J. B. Sidgwick, Faber & Faber.

MATHEMATICAL ASTRONOMY
Elements of Mathematical Astronomy, M. Davidson, Hutchinson.
Foundations of Astronomy, W. M. Smart, Longmans.

INDISPENSABLE GUIDES FOR THE AMATEUR OBSERVER
Star Atlas, Norton, Gall & Inglis.
The Amateur Astronomer's Handbook, and *Observational Astronomy for Amateurs* J. B. Sidgwick, Faber & Faber.

PHYSICS OF THE STARS
From Atoms to Stars, M. Davidson, Hutchinson.
Frontiers of Astronomy, David S. Evans, Sigma Books.
Astronomy of Stellar Energy and Decay, Martin Johnson, Faber and Faber.
Frontiers of Astronomy, F. Hoyle, Heinemann.

AN IMPORTANT SERIES:
Harvard Books on Astronomy, Blakiston Company.
Atoms, Stars and Nebulae, Goldberg and Aller.
The Story of Variable Stars, Campbell and Jacchia.
Our Sun, Donald H. Menzel.
Galaxies, H. Shapley.
The Milky Way, Bok and Bok.

TELESCOPE MAKING:
Amateur Telescope Making and *Amateur Telescope Making, Advanced*, Scientific American Publishing Company.
Telescopes and Accessories, Dimitroff and Baker (Harvard Books on Astronomy).

Eclipses

Eclipses of the Sun and Moon, Dyson and Woolley, Oxford University Press.

Planets

The Planet Mars, de Vaucouleurs, Faber and Faber.

Life on Other Worlds, Sir Harold Spencer Jones, English Universities Press.

Earth, Moon and Planets, Fred L. Whipple, (Harvard Books on Astronomy).

Guide to the Planets, Patrick Moore, Eyre & Spottiswoode.

Guide to the Moon, Patrick Moore, Eyre & Spottiswoode.

Comets

Comets and Meteor Streams, J. G. Porter, Chapman and Hall.

Between the Planets, F. G. Watson, (Harvard Books on Astronomy).

Radio Astronomy, Lovell and Clegg, Chapman and Hall.

An Important Reference Book (rather advanced)

Astrophysical Quantities, C. W. Allen, Athlone Press.

Atmospheric Phenomena

A fascinating book is *Light and Colour in the Open Air*, Minnaert, G. Bell & Sons.

Journals

The Observatory Magazine, c/o Royal Greenwich Observatory, Herstmonceux Castle, Hailsham, Sussex, England (fairly technical: reports of Royal Astronomical Society meetings, correspondence, etc.).

Sky and Telescope, Harvard College Observatory, Cambridge, Mass., U.S.A. (up-to-date, profusely illustrated, fairly simple presentation).

Astronomical articles frequently appear in *Discovery*, *Endeavour*, and *Penguin Science News*.

ASTRONOMICAL ORGANISATIONS

The following bodies promote interest in astronomy among amateurs and will probably be glad to offer advice and help. Most of them publish journals.

GREAT BRITAIN

British Astronomical Association, 303, Bath Road, Hounslow West, Middlesex, England.

U.S.A.

Numerous astronomical societies exist. To make contact with the nearest try,

Sky and Telescope, Harvard College Observatory, 60, Garden Street, Cambridge 38, Mass., U.S.A. or Astronomical Society of the Pacific, 675, Eighteenth Avenue, San Francisco 21, California, U.S.A.

CANADA

Royal Astronomical Society of Canada, 13 Ross Street, Toronto 2B, Canada.

IRELAND

The Irish Astronomical Society; c/o Armagh Observatory, Armagh, Northern Ireland.

NEW ZEALAND

Royal Astronomical Society of New Zealand, Carter Observatory, Wellington, N.Z.

SOUTH AFRICA

Astronomical Society of South Africa, Royal Observatory, Observatory, Cape Province, Union of South Africa.

AUSTRALIA

Several astronomical societies exist: enquire of Commonwealth Astronomer, Mount Stromlo, Canberra.

INDEX